ENERGY SCIENCE, ENGINEERING AND TECHNOLOGY

FEDERAL INCENTIVES FOR RENEWABLE ENERGY

Energy Science, Engineering and Technology

Additional books in this series can be found on Nova's website under the Series tab.

Additional E-books in this series can be found on Nova's website under the E-books tab.

ENERGY SCIENCE, ENGINEERING AND TECHNOLOGY

FEDERAL INCENTIVES FOR RENEWABLE ENERGY

LIAM BELISARIO
AND
NIKLAS MEIER
EDITORS

Nova Science Publishers, Inc.
New York

For permission to use material from this book please contact us:
Telephone 631-231-7269; Fax 631-231-8175
Web Site: http://www.novapublishers.com

Library of Congress Cataloging-in-Publication Data

Federal incentives for renewable energy / editors, Liam Belisario and Niklas Meier.
p. cm.
Includes index.
ISBN 978-1-61324-946-8 (hardcover)
1. Energy policy--United States. 2. Renewable energy sources--United States. I. Belisario, Liam. II. Meier, Niklas. HD9502.U52F434 2011
333.79'41580973--dc23

2011018571

Published by Nova Science Publishers, Inc. † New York

CONTENTS

PREFACE

Energy is crucial to the operation of a modern industrial and service economy. Recently, there have been growing concerns about the availability and cost of energy, as well as the environmental impacts of fossil energy use, especially global climate change. Those combined concerns have rekindled interest in energy efficiency, energy conservation and the development and commercialization of renewable energy technologies. This book examines federal programs that provide grants, loans, loan guarantees and other direct or indirect regulatory incentives for energy efficiency, energy conservation and renewable energy.

Chapter 1- Energy is crucial to the operation of a modern industrial and services economy. Recently, there have been growing concerns about the availability and cost of energy and about environmental impacts of fossil energy use, especially global climate change. Those combined concerns have rekindled interest in energy efficiency, energy conservation, and the development and commercialization of renewable energy technologies.

Many of the existing efficiency and renewables programs have authorizations tracing back to the 1970s. Many of the programs have been reauthorized and redesigned repeatedly to meet changing economic factors. The programs apply broadly to sectors ranging from industry to academia, and from state and local governments to rural communities.

Chapter 2- The following list of authoritative resources is designed to assist in responding to a broad range of constituent questions and concerns about renewable energy and energy efficiency tax incentives. Links are provided for the following: the full text of public laws establishing and extending federal renewable energy and energy efficiency incentives; federal, state, and local incentives resources; incentive resources grouped by technology type (solar, wind, geothermal, and biomass); and federal grants information resources. The last section of this report includes tables displaying popular incentives, the corresponding *U.S. Code* citations, and current expiration dates of those incentives.

Chapter 3- Congress created the Section 1603 grant program as part of the American Recovery and Reinvestment Act of 2009 (ARRA; P.L. 111-5). This program, administered by the U.S. Department of the Treasury, provides cash grant incentives for renewable energy projects. Initially, the Section 1603 grant program was scheduled to expire at the end of 2010. A one-year extension was enacted as part of the Tax Relief, Unemployment Reauthorization, and Job Creation Act of 2010 (P.L. 111-312) at an estimated cost of $3 billion. Absent Congressional action, the Section 1603 grant program will expire at the end of 2011.

Chapter 4- With recent high energy prices, the passage of major energy legislation in 2005 (P.L. 109-58) and 2007 (P.L. 110-140), and the passage of a farm bill in 2008 (P.L. 110-246), there is ongoing congressional interest in promoting alternatives to petroleum fuels. Biofuels—transportation fuels produced from plants and other organic materials—are of particular interest.

Ethanol and biodiesel, the two most widely used biofuels, receive significant government support under federal law in the form of mandated fuel use, tax incentives, loan and grant programs, and certain regulatory requirements. The 22 programs and provisions listed in this report have been established over the past three decades, and are administered by five separate agencies and departments: Environmental Protection Agency, U.S. Department of Agriculture, Department of Energy, Internal Revenue Service, and Customs and Border Protection. These programs target a variety of beneficiaries, including farmers and rural small businesses, biofuel producers, petroleum suppliers, and fuel marketers. Arguably, in prior years the most significant federal programs for biofuels have been tax credits for the production or sale of ethanol and biodiesel. However, with the establishment of the renewable fuel standard (RFS) under P.L. 109-58, Congress has mandated biofuels use; P.L. 110-140 significantly expanded that mandate. In the long term, the mandate may prove even more significant than tax incentives in promoting the use of these fuels.

In: Federal Incentives for Renewable Energy ISBN: 978-1-61324-946-8
Editors: L.Belisario and N. Meier

Chapter 1

RENEWABLE ENERGY AND ENERGY EFFICIENCY INCENTIVES: A SUMMARY OF FEDERAL PROGRAMS*

Lynn J. Cunningham and Beth A. Roberts

SUMMARY

Energy is crucial to the operation of a modern industrial and services economy. Recently, there have been growing concerns about the availability and cost of energy and about environmental impacts of fossil energy use, especially global climate change. Those combined concerns have rekindled interest in energy efficiency, energy conservation, and the development and commercialization of renewable energy technologies.

Many of the existing efficiency and renewables programs have authorizations tracing back to the 1970s. Many of the programs have been reauthorized and redesigned repeatedly to meet changing economic factors. The programs apply broadly to sectors ranging from industry to academia, and from state and local governments to rural communities.

Since 2005, Congress has enacted several major energy laws: the Energy Policy Act of 2005 (EPACT 2005; P.L. 109-58); the Energy

---

* This is an edited, reformatted and augmented version of a Congressional Research Service publication, CRS Report for Congress R40913, from www.crs.gov, dated March 22, 2011.

Independence and Security Act of 2007 (EISA; P.L. 110- 140); the Energy Improvement and Extension Act (EIEA), enacted as Division B of the Emergency Economic Stabilization Act (EESA; P.L. 110-343); and the American Reinvestment and Recovery Act (ARRA; P.L. 111-5). Each of those laws established, expanded, or modified energy efficiency and renewable energy research, development, demonstration, and deployment (RDD&D) programs. The Department of Energy (DOE) operates the greatest number of efficiency and renewable energy incentive programs. The Department of the Treasury and the Department of Agriculture (USDA) operate several programs. A few programs can also be found among the Departments of Transportation (DOT), Labor, and Housing and Urban Development (HUD).

This report describes federal programs that provide grants, loans, loan guarantees, and other direct or indirect regulatory incentives for energy efficiency, energy conservation, and renewable energy. For each program, the report provides the administering agency, authorizing statute(s), annual funding, and the program expiration date. The appendixes provide summary information in a tabular format.

INTRODUCTION

The United States has an abundance of natural resources. For much of the nation's history, energy was not a concern as commerce and industry needs could be met by domestic supplies. However, industrialization and population growth, and the continuing development of a consumer-oriented society, soon led to the necessity of obtaining foreign sources of energy to supplement the demands of a growing economy.

Recognition of the implications of dependence on foreign sources of energy, coupled with concerns over volatility of prices driven by fluctuations in supply spurred by world events, have led to efforts to increase U.S. energy independence and reduce domestic consumption. The result has been the emergence of a number of programs focused on energy efficiency and conservation of domestic resources and on research programs that target the development of renewable sources of energy. Many of these programs have roots going back almost 40 years and have been redesigned many times over that period.

Many of the current programs have been reauthorized and redesigned periodically to meet changing economic conditions and national interests. The programs apply broadly to sectors ranging from industry to academia, and from state and local governments to rural communities. Each program has been designed to meet current needs as well as future anticipated challenges.

Since 2005, Congress has enacted several major energy laws: the Energy Policy Act of 2005 (EPACT 2005; P.L. 109-58); the Energy Independence and Security Act of 2007 (EISA, P.L. 110- 140); the Energy Improvement and Extension Act (EIEA), enacted as Division B of the Emergency Economic Stabilization Act (EESA, P.L. 110-343); and the American Reinvestment and Recovery Act (ARRA, P.L. 111-5). Each of those laws established, expanded, or modified energy efficiency and renewable energy research, development, demonstration, and deployment (RDD&D) programs. The Department of Energy (DOE) operates the greatest number of efficiency and renewable energy incentive programs. The Department of the Treasury and the Department of Agriculture (USDA) operate several programs. A few programs can also be found among the Departments of Transportation (DOT), Labor, and Housing and Urban Development (HUD).

This report outlines current federal programs and provisions providing grants, loan, loan guarantees, and other direct or indirect incentives for energy efficiency, energy conservation, and renewable energy RD&D. The programs are grouped by administering agency with information on links to applicable federal agency websites. Incentives are summarized and indexed in the appendixes.

ARRA funding opportunities are available to eligible energy projects with start dates generally in 2009 and 2010. Applicants should check the DOE's ARRA website for specific application deadline details.

The program descriptions were compiled from authorizing statutes, the U.S. Code, and Administration budget request documents. Program descriptions and some funding information were compiled from The Database of State Incentives for Renewables and Efficiency (DSIRE), the *Catalog of Federal Domestic Assistance* (CFDA) and the Energy Star website. Most budgetary figures were compiled from executive agency budget justifications and congressional committee reports.

I. Department of Energy Office of Energy Efficiency and Renewable Energy

Renewable Energy

Biomass

1. Biomass and Biorefinery Systems R&D Program

- Administered by: Office of Energy Efficiency and Renewable Energy (EERE)
- Authorization: Federal Nonnuclear Energy Research and Development Act of 1974 (P.L. 93-577); Energy Policy and Conservation Act of 1975 (EPCA; P.L. 94-163); Energy Conservation and Production Act of 1976 (ECPA; P.L. 94- 385); Department of Energy Organization Act of 1977 (P.L. 95-91); Energy Tax Act (P.L. 95-618); National Energy Conservation Policy Act of 1978 (NECPA; P.L. 95-619); Powerplants and Industrial Fuel Use Act of 1978 (P.L. 95-620); Energy Security Act of 1980 (P.L. 96-294); National Appliance Energy Conservation Act of 1987 (P.L. 100-12); Federal Energy Management Improvement Act of 1988 (P.L. 100-615); Renewable Energy and Energy Efficiency Technology Competitiveness Act of 1989 (P.L. 101-218); Clean Air Act Amendments of 1990 (P.L. 101-549); Solar, Wind, Waste, and Geothermal Power Production Incentives Act of 1990 (P.L. 101-575); Energy Policy Act of 1992 (EPACT; P.L. 102-486); Biomass Research and Development Act of 2000 (Title III of Agricultural Risk Protection Act of 2000; P.L. 106-224); Farm Security and Rural Investment Act of 2002 (P.L. 107-171); Healthy Forest Restoration Act of 2003 (P.L. 108-148); Energy Policy Act of 2005 (EPACT 2005; P.L. 109-58); Energy Independence and Security Act of 2007 (EISA; P.L. 110-140); The Food, Conservation, and Energy Act of 2008 (P.L. 110-234); American Recovery and Reinvestment Act of 2009 (ARRA; P.L. 111-5)
- Annual funding: \$89.8 million for FY2006; \$196.3 million for FY2007; \$195.6 million for FY2008; \$214 million for FY2009; an additional \$777 million in FY2009 from ARRA; \$220 million for FY2010; and \$220 million requested for FY2011
- Scheduled termination: None

- Description: This program works with industrial partners, national laboratories, universities, and other stakeholders to develop the technologies and systems needed to cost-effectively transform the nation's renewable and abundant domestic biomass resources into clean, affordable, and sustainable biofuels, bioproducts, and biopower. In recent years, the program has been primarily geared toward development and deployment of ethanol from non-food feedstocks, but is now expanding its scope to additional alternative fuels, such as bio-butanol, green gasoline, jet fuel, and diesel.
- Qualified applicant: Colleges and universities; profit organizations
- Qualified technologies: Biomass

2. Regional Biomass Energy Grant Programs

- Administered by: Office of Biomass Program, EERE
- Authorization: Department of Energy Organization Act of 1977 (P.L. 95-91); Energy and Water Development Appropriations Act for FY1987 (P.L. 99-500)
- Annual funding: $395,000 for FY2007; an estimated $75,131 for FY2008; an estimated $25,705 for FY2009; an estimated $4.8 million for FY2010; data for FY2011 is currently unavailable
- Scheduled termination: None
- Description: This program provides assistance to increase America's use of fuels, chemicals, materials, and power made from domestic biomass on a sustainable basis. Assistance may be used to develop and transfer any of several biomass energy technologies to the scientific and industrial communities. For regional programs, such technologies will be appropriate for the needs and resources of particular regions of the United States.
- Qualified applicants: State and local governments; colleges and universities; profit organizations; nonprofit organizations
- Qualified technologies: Biomass
- For more information: See CFDA program number 81.079 at https://www.cfda.gov/

Geothermal

3. Geothermal Technologies Program (GTP)

- Administered by: EERE
- Authorization: Geothermal Energy Research, Development, and Demonstration Act (P.L. 93-410); Department of Energy Organization Act (P.L. 95-91); Energy Tax Act of 1978 (P.L. 95-618); Energy Security Act of 1980 (P.L. 96-294); Renewable Energy and Energy Efficiency Technology Competitiveness Act of 1989 (P.L. 101-218); Solar, Wind, Waste, and Geothermal Power Production Incentives Act of 1990 (P.L. 101-575); Energy Policy Act of 1992 (EPACT; P.L. 102-486); Energy Policy Act of 2005 (EPACT 2005; P.L. 109-58); Energy Independence and Security Act of 2007 (EISA; P.L. 110-140); American Recovery and Reinvestment Act of 2009 (ARRA; P.L. 111-5)
- Annual funding: $68.2 million for FY2006; $5 million for FY2007; $19.3 million for FY2008; $43.3 million for FY2009; an additional $393 million appropriated in FY2009 from ARRA; $44 million for FY2010; and $55 million requested for FY2011
- Scheduled termination: None
- Description: This program partners the federal government with industry, academia, and research facilities to further the development of geothermal energy technologies. Competitive solicitations issued as Funding Opportunity Announcements (FOAs) are the principal mechanism used by the GTP to contract for cost-shared research, development, and demonstration projects.
- Qualified applicants: Profit organizations; colleges and universities
- Qualified technology: Geothermal
- For more information: See EERE's Geothermal Technologies Program Website at http://www1.eere.energy and CFDA Program number 81.087 at https://www.cfda.gov/

Hydrogen and Fuel Cells

4. Hydrogen & Fuel Cell Technologies Program

- Administered by: EERE
- Authorization: Federal Energy Administration Act of 1974 (P.L. 93-275); Federal Nonnuclear Energy Research and Development Act of 1974 (P.L. 93- 577); Energy Policy and Conservation Act of 1975

(EPCA; P.L. 94-163); Electric and Hybrid Vehicle Research, Development and Demonstration Act (P.L. 94-413); Department of Energy Organization Act of 1977 (P.L. 95-91); Automotive Propulsion Research and Development Act of 1978 (Title III of Department of Energy Act of 1978-Civilian Applications; P.L. 95-238); Methane Transportation Research, Development and Demonstration Act of 1980 (P.L. 96-512); Energy Security Act of 1980 (P.L. 96-294); Alternative Motor Fuels Act of 1988 (P.L. 100-494); Spark M. Matsunaga Hydrogen Research, Development, and Demonstration Act of 1990 (P.L. 101-566); Energy Policy Act of 1992 (EPACT; P.L. 102-486); Hydrogen Future Act of 1996 (P.L. 104-271); Energy Policy Act of 2005 (EPACT 2005; P.L. 109-58); Energy Independence and Security Act of 2007 (EISA; P.L. 110-140); American Recovery and Reinvestment Act of 2009 (ARRA; P.L. 111-5)

- Annual funding: $153.4 million for FY2006; $190 million for FY2007; $206.2 million for FY2008; $164.6 million for FY2009; an additional $43 million appropriated in FY2009 from ARRA; $174 million for FY2010; and $137 million requested for FY2011
- Scheduled termination: None
- Description: This program partners with industry, academia, and national laboratories and works in close coordination with Vehicle Technologies and other programs at DOE to: overcome technical barriers through R&D of hydrogen production, delivery, and storage technologies; overcome technical barriers to fuel cell technologies for transportation, distributed stationary power, and portable power applications; address safety issues and facilitate the development of model codes and standards; validate and demonstrate hydrogen and fuel cells in real-world conditions; and educate key stakeholders whose acceptance of these technologies will determine their success in the marketplace.
- Qualified applicants: Federal government; national laboratories; colleges and universities; and profit organizations
- Qualified technologies: Hydrogen and fuel cells
- For more information: See EERE's Hydrogen and Fuel Cell Technologies Website at http://www1.eere.energy.gov/ hydrogenand fuelcells/and CFDA Program number 81.087 at https:// www. fda.gov/

Solar

5. Solar Energy Technologies Program (SETP)

- Administered by: EERE
- Authorization: Energy Policy and Conservation Act of 1975 (EPCA; P.L. 94- 163); Energy Conservation and Production Act of 1976 (ECPA; P.L. 94-385); Department of Energy Organization Act of 1977 (P.L. 95-91); Solar Photovoltaic Energy Research, Development and Demonstration Act of 1984 (P.L. 95-590); National Energy Conservation Policy Act of 1978 (NECPA; P.L. 95-619); Energy Security Act of 1980 (P.L. 96-294); Renewable Energy and Energy Efficiency Technology Competitiveness Act of 1989 (P.L. 101-218); Solar, Wind, Waste, and Geothermal Power Production Incentives Act of 1990 (P.L. 101-575); Solar, Wind, Waste, and Geothermal Power Production Incentives Technical Amendments Act of 1991 (P.L. 102-46); Energy Policy Act of 1992 (EPACT; P.L. 102-486); Energy Policy Act of 2005 (EPACT 2005; P.L. 109-58); Energy Independence and Security Act of 2007 (EISA; P.L. 110-140); American Recovery and Reinvestment Act of 2009 (ARRA; P.L. 111-5)
- Annual funding: $81.8 million for FY2006; $157 million for FY2007; $166.3 million for FY2008; $172.4 million for FY2009; an additional $116 million appropriated in FY2009 from ARRA; $247 million for FY2010; and $302.4 million requested for FY2011
- Scheduled termination: None
- Description: SETP partners with industry, national laboratories, and universities to develop and bring reliable and affordable solar energy technologies to the marketplace. This program finances R&D in four major subprograms: Photovoltaics (PV); Concentrating Solar Power (CSP); Systems Integration for Solar Technologies; and Market Transformation for Solar Technologies.
- Qualified applicants: Industry; national laboratories; colleges and universities
- Qualified technology: Solar
- For more information: See EERE's Solar Energy Technologies Program Website http://www1.eere.energy and CFDA Program number 81.087 at https://www.cfda.gov/

Wind and Hydropower

6. Wind and Hydropower Technologies Program

- Administered by: EERE
- Authorization: Energy Policy and Conservation Act of 1975 (EPCA; P.L. 94- 163); Renewable Energy and Energy Efficiency Technology Competitiveness Act of 1989 (P.L. 101-218); Solar, Wind, Waste, and Geothermal Power Production Incentives Act of 1990 (P.L. 101-575); Energy Policy Act of 1992 (EPACT; P.L. 102-486); Energy Policy Act of 2005 (EPACT 2005; P.L. 109- 58); Energy Independence and Security Act of 2007 (EISA; P.L. 110-140); American Recovery and Reinvestment Act of 2009 (ARRA; P.L. 111-5)
- Annual funding: $39.8 million for FY2006 ($38.3 million for wind; $495,000 for hydropower); $48.7 million for FY2007 ($48.7 million for wind; $0 for hydropower); $58.7 million for FY2008 ($49 million for wind; $9.7 million for hydropower); $93.5 million for FY2009 ($54.4 million for wind; $39.1 million for hydropower); an additional $138.6 million appropriated in FY2009 from ARRA ($106.9 million for wind; $31.7 million for hydropower); $130 million for FY2010 ($80 million for wind; $50 million for hydropower); and $163 million requested for FY2011 ($122.5 million for wind; $40.5 million for hydropower)
- Scheduled termination: None
- Description: The program partners with federal, state, and other stakeholder groups to conduct research and development activities through competitively selected, cost-shared research and development projects with industry to improve the performance, lower the costs, and accelerate the deployment of wind and water power technologies.
- Qualified applicants: Federal, state, local, and tribal governments; national laboratories; industry; small businesses; colleges and universities
- Qualified technologies: Wind; hydroelectric; hydrokinetic energy; wave energy; tidal energy; ocean thermal energy conversion
- For more information: See EERE's Wind and Hydropower Technologies Program website at
- http://www1.eere.energy
- For more information, see CFDA Program number 81.087 at https://www.cfda.gov/

Energy Efficiency

Buildings

7. Building Technologies Program

- Administered by: EERE
- Authorization: Energy Policy and Conservation Act of 1975 (EPCA; P.L. 94- 163); Energy Conservation and Production Act of 1976 (ECPA; P.L. 94-385); Department of Energy Organization Act of 1977 (P.L. 95-91); Energy Tax Act of 1978 (P.L. 95-618); National Energy Conservation Policy Act of 1978 (NECPA; P.L. 95-619); Power Plant and Industrial Fuel Use Act of 1978 (P.L. 95-620); Energy Security Act (P.L. 96-294); National Appliance Energy Supply Act of 1987 (P.L. 100-12); National Appliance Energy Conservation Amendments of 1988 (P.L. 100-357); Federal Energy Management Improvement Act of 1988 (P.L. 100-615); Energy Policy Act of 1992 (EPACT; P.L. 102-486); Energy Policy Act of 2005 (EPACT 2005; P.L. 109-58); Energy Independence and Security Act of 2007 (EISA; P.L. 110-140); American Recovery and Reinvestment Act of 2009 (ARRA; P.L. 111-5)
- Annual funding: $68.2 million for FY2006; $103 million for FY2007; $107.4 million for FY2008; $138.1 million for FY2009; an additional $319.2 million appropriated in FY2009 from ARRA; $222 million for FY2010; and $230.7 million requested for FY2011
- Scheduled termination: None
- Description: In partnership with the private sector, state and local governments, national laboratories, and universities, the Building Technologies Program works to improve the efficiency of buildings and the equipment, components, and systems within them. The program supports research and development (R&D) activities and provides tools, guidelines, training, and access to technical and financial resources.
- Qualified applicants: State and local governments; universities; national laboratories
- Qualified technologies: Energy-efficient innovations for building envelopes, equipment, lighting, daylighting, and windows; passive solar; photovoltaics; fuel cells; advanced sensors and controls; and combined heating, cooling, and power systems

- For more information: See EERE's Building Technologies Program website at http://www1.eere.energy

8. Energy Efficient Appliance Rebate Program (EEARP)

- Administered by: EERE
- Authorization: Energy Policy Act of 2005 (EPACT 2005; P.L. 109-58), Title I, Part B; American Recovery and Reinvestment Act of 2009 (ARRA; P.L. 111-5)
- Annual funding: $0 for FY2008; $298.5 million in FY2009 from ARRA; $0 for FY2010; $0 requested for FY2011
- Scheduled termination: None
- Description: The program provides financial and technical assistance to states to establish residential Energy Star rated appliance rebate programs. The program's objectives are: to reduce fossil fuel emissions created as a result of activities within the jurisdictions of eligible entities; and to improve energy efficiency in the residential sector.
- Qualified applicants: State governments, including U.S territories and possessions
- Qualified technologies: Energy efficient appliances
- For more information: See CFDA Program number 81.127 at https://www.cfda.gov

9. Weatherization Assistance Program (WAP)

- Administered by: EERE
- Authorization: Energy Supply and Production Act of 1976 (ECPA; P.L. 94-385); National Energy Conservation Policy Act of 1978 (NECPA; P.L. 95-619); Energy Security Act of 1980 (P.L. 96-294); Energy Policy Act of 1992 (EPACT; P.L. 102-486); Energy Policy Act of 2005 (EPACT 2005; P.L. 109-58); Energy Independence and Security Act of 2007 (EISA; P.L. 110-140); American Recovery and Reinvestment Act of 2009 (ARRA; P.L. 111-5)
- Annual funding: $227.2 million for FY2008; $450 million for FY2009; an additional $5 billion appropriated in FY2009 from ARRA; $270 million for FY2010; and $385 million requested for FY2011
- Scheduled termination: None
- Description: This program reduces energy costs for low-income households by increasing the energy efficiency of their homes while

ensuring their health and safety. DOE provides funding and technical guidance to states, which manage the day-to-day details of the program. Low-income families receive services from a network of more than 900 local weatherization service providers who install energy efficiency measures in the homes of qualifying homeowners free of charge.

- Qualified applicants: State and tribal governments, including U.S. territories
- Qualified technologies: Weatherization technologies include a wide range of energy efficiency measures for retrofitting homes and apartment buildings. Weatherization service providers choose the best package of efficiency measures for each home based on an energy audit of the home. Typical measures may include installing insulation, sealing ducts, tuning and repairing heating and cooling systems and if indicated, replacement of the same; mitigating air infiltration; and reducing electric base load consumption. For more information: See EERE's Weatherization Assistance Program website at http://apps1.eere.energy.gov/weatherization/; and CFDA program number 81.042 at https://www.cfda.gov

Industrial

10. Inventions and Innovations Program

- Administered by: EERE
- Authorization: Federal Nonnuclear Energy Research and Development Policy Act (P.L. 93-577), Section 14; 42 USC 5913
- Annual funding: $2.8 million for FY2007; $145,000 for FY2008; $1.8 million for FY2009; an estimated $102,000 for FY2010; and an estimated $150,000 for FY2011
- Scheduled termination: None
- Description: This program provides financial and technical assistance for research and development of innovative, energy-saving ideas and inventions with future commercial market potential. Inventions and Innovations support energy efficiency and renewable energy technology development in focus areas that align with Office of Energy Efficiency and Renewable Energy programs.
- Qualified applicants: Individuals; small businesses
- Qualified technologies: Specific energy efficiency and renewable energy technologies not listed

- For more information: See CFDA Program number 81.036 at https://www.cfda/ gov. The U.S. Department of Energy's Inventions & Innovations website has been retired. To access information on financial opportunities and current solicitations, visit the Industrial Technologies Program site at http://www1.eere.energy.gov/industry/financial/index.html

11. Industrial Technologies Program (ITP)

- Administered by: EERE
- Authorization: Energy Policy and Conservation Act of 1975 (EPCA; P.L. 94- 163); Energy Conservation and Production Act of 1976 (ECPA; P.L. 94-385); Department of Energy Organization Act of 1977 (P.L. 95-91); National Energy Conservation Policy Act of 1978 (NECPA; P.L. 95-619); Powerplant and Industrial Fuel Use Act of 1978 (P.L. 95-620); Energy Security Act of 1980 (P.L. 96-294); Renewable Energy and Energy Efficiency Technology Competitiveness Act of 1989 (P.L. 101-218); Energy Policy Act of 1992 (EPACT; P.L. 102-486); Energy Policy Act of 2005 (EPACT 2005; P.L. 109-58); Energy Independence and Security Act of 2007 (EISA; P.L. 110-140); American Recovery and Reinvestment Act of 2009 (ARRA; P.L. 111-5)
- Annual funding: $55.9 million for FY2006; $55.8 million for FY2007; $63.2 million for FY2008; $88.2 million for FY2009; an additional $261.5 million appropriated in FY2009 from ARRA; $96 million for FY2010; and $100 million requested for FY2011
- Scheduled termination: None
- Description: ITP works with industry to improve industrial energy efficiency and environmental performance while increasing productivity conducting R&D on new energy efficient technologies; supporting commercialization of emerging technologies; providing plants with access to proven technologies, energy assessments, software tools, and other resources; and promoting energy and carbon management in industry.
- Qualified applicants: Industrial organizations
- Qualified technologies: Crosscutting technologies that improve the efficiency of technologies that are common to many industrial processes and can benefit multiple industries. Crosscutting technology R&D areas include combustion; distributed energy; energy intensity

processes; fuel and feedstock liability; industrial materials for the future; nanomanufacturing; and sensors and automation

- For more information: See EERE's Industrial Technologies Program website at http://www1.eere.energy.gov/industry/

Vehicles

12. Vehicle Technologies Program

- Administered by: EERE
- Authorization: Department of Energy Organization Act of 1977 (P.L. 95-91); Energy Policy Act of 1992 (EPACT; P.L. 102-486); Energy Policy Act of 2005 (EPACT 2005; P.L. 109-58); Energy Independence and Security Act of 2007 (EISA; P.L. 110-140); American Recovery and Reinvestment Act of 2009 (ARRA; P.L. 111-5)
- Annual funding: $178.4 million for FY2006; $183.6 million for FY2007; $208.4 million for FY2008; $267.1 million for FY2009; an additional $2.8 billion appropriated in FY2009 from ARRA; $311.4 million for FY2010; and $325.3 million requested for FY2011
- Scheduled termination: None
- Description: The Vehicle Technologies Program works with industry leaders to develop and deploy advanced transportation technologies that could achieve significant improvements in vehicle fuel efficiency and displace oil with other fuels that ultimately can be domestically produced in a clean and cost-competitive manner. Program activities include research, development, demonstration, testing, technology validation, technology transfer, and education.
- Qualified applicants: Industry; colleges and universities; federal, state and local governments; national laboratories
- Qualified technologies: Hybrid electric systems; biofuels or fuels technology; advanced internal combustion engines; advanced propulsion materials
- For more information: See EERE's Vehicle Technology Program website at http://www1.eere.energy and Vehicle Technologies Program Factsheet at
- http://www1.eere.energy vehicles_fs.pdf

Other Energy Efficiency and Renewable Energy Programs

13. Conservation Research and Development Grants

- Administered by: EERE
- Authorization: Federal Nonnuclear Energy Research and Development Act of 1974 (P.L. 93-577); Department of Energy Organization Act of 1977 (P.L. 95- 91); Continuing Appropriations Act for FY1983 (P.L. 97-377); American Recovery and Reinvestment Act of 2009 (ARRA; P.L. 111-5)
- Annual funding: $87.5 million for FY2008; $203.7 million for FY2009; an estimated $1.7 billion for FY2010; and an estimated $180 million for FY2011
- Scheduled termination: None
- Description: This program provides project grants to conduct balanced, longterm research efforts in buildings technologies, industrial technologies, vehicle technologies, and hydrogen and fuel cell technologies.
- Qualified applicant: State, local, and tribal governments; universities; profit organizations; and private nonprofit institutions/organizations
- Qualified technologies: Hydrogen and fuel cells; energy efficient technologies; advanced battery manufacturing
- For more information: See CFDA program number 81.086 at https://www.cfda.gov/

14. Energy Efficiency and Conservation Block Grants Program (EECBG)

- Administered by: EERE
- Authorization: Energy Independence and Security Act of 2007 (EISA; P.L.110-140), Title V, Subtitle E; American Recovery and Reinvestment Act of 2009 (ARRA; P.L. 111-5)
- Annual funding: $0 for FY2008; $3.2 billion for FY2009 from ARRA; $0 for FY2010; and $0 requested for FY2011
- Scheduled termination: None
- Description: This program is part of DOE's Weather and Intergovernmental Program. The EECBG Program provides formula and competitive grants to empower local communities to make strategic investments to meet the nation's long-term goals for energy independence and leadership on climate change. Grants can be used for energy efficiency and conservation programs and projects

community-wide, as well as renewable energy installations on government buildings.

- Qualified applicants: State, local, and tribal governments, including U.S. territories
- Qualified technologies: Energy efficient equipment and lighting; combined heating and cooling systems; combined heat and power systems; solar; wind; fuel cells; biomass
- For more information: See EERE's Energy Efficiency and Conservation Block Grants Program Website http://www.eecbg.energy and CFDA program number 81.128 at https://www.cfda.gov

15. Energy Efficiency and Renewable Energy Information Dissemination, Outreach, Training, and Technical Analysis/Assistance Grant Program

- Administered by: EERE
- Authorization: Energy Reorganization Act of 1974 (P.L. 93-438); Department of Energy Organization Act of 1977 (P.L. 95-91); Energy Policy Act of 1992 (EPACT; P.L. 102-486)
- Annual funding: $30 million for FY2007; $39.7 million for FY2008; $38 million for FY2009; an estimated $41.2 million for FY2010; and an estimated $45 million for FY2011
- Scheduled termination: None
- Description: This program provides financial assistance for information dissemination, outreach, training and related technical analysis/assistance which will: (1) stimulate increased energy efficiency in transportation, buildings, industry and the Federal sector and encourage increased use of renewable and alternative energy; and (2) accelerate the adoption of new technologies to increase energy efficiency and the use of renewable and alternative energy through the competitive solicitation of applications.
- Qualified applicants: State and local governments; Native American organizations; individuals; universities; profit organizations; private nonprofit organizations; public nonprofit organizations; and Alaskan Native corporations
- Qualified technologies: Specific energy efficiency and renewable energy technologies not listed
- For more information: See CFDA Program number 81.117 at https://www.cfda.gov/

16. Energy Efficiency and Renewable Energy Technology Deployment, Demonstration, and Commercialization Grant Program

- Administered by: EERE
- Authorization: Energy Policy Act of 1992 (EPACT; P.L. 102-486); Energy Policy Act of 2005 (EPACT 2005; P.L. 109-58); Energy Independence and Security Act of 2007 (EISA; P.L. 110-140); American Recovery and Reinvestment Act of 2009 (ARRA; P.L. 111-5)
- Annual Funding: $0 for FY2008; $21.8 million for FY2009; an estimated $7.2 million for FY2010. It is anticipated that all funds obligated under this program in FY2010 will be Recovery Act funds. FY2011 budget unconfirmed to date
- Scheduled termination: None
- Description: This program provides financial assistance for the technology deployment, demonstration, and commercialization of energy efficiency and renewable energy technologies. This includes biomass, building technologies, federal energy management, geothermal technologies, projects involving hydrogen, fuel cells and infrastructure technologies, industrial technologies, solar energy technologies, vehicle technologies, weatherization and inter-governmental, and wind and hydropower technologies.
- Qualified applicants: State governments; profit organizations
- Qualified technologies: Biomass; geothermal; hydrogen and fuel cell technologies; solar; hydropower
- For more information: See CFDA Program number 81.129 at https://www.cfda.gov
-

17. Renewable Energy Production Incentive (REPI)Administered by: EERE

- Authorization: Energy Policy Act of 1992 (EPACT; P.L. 102-486), Title XII, Section 1212; Energy Policy Act of 2005 (EPACT 2005; P.L. 109-58), Title II, Subtitle A, Section 202); 42 USC 13317
- Annual funding: $4.95 million for FY2006; $4.95 million for FY2007; $4.95 million for FY2008; $5 million for FY2009; $0 for FY2010; $0 requested for FY2011
- Scheduled termination: End of FY2026
- Description: This program provides incentive payments for electricity generated and sold by new qualifying renewable energy facilities. Qualifying systems are eligible for annual incentive payments of 1.5¢

per kilowatt-hour in 1993 dollars (indexed for inflation) for the first 10-year period of their operation, subject to the availability of annual appropriations in each federal fiscal year of operation.

- Qualified applicants: State, local, and tribal governments; public utilities; not-for-profit electrical cooperatives; Native American corporations
- Qualified technologies: Solar thermal electric; photovoltaics; landfill gas; wind; biomass; geothermal electric; anaerobic digestion; tidal energy; wave energy; ocean thermal
- For more information: See EERE's Renewable Energy Production Incentive Program website at http://apps1.eere.energy and DSIRE at http://www.dsireusa.org/incentives/incentive.cfm? Incentive_Code=US33F&re= 1&ee=1

18. Renewable Energy Research and Development Program

- Administered by: EERE
- Authorization: Department of Energy Organization Act of 1977 (P.L. 95-91); Department of Energy Act of 1978 - Civilian Applications (P.L. 95-238), Section 207; Renewable Energy and Energy Efficiency Technology Competitiveness Act of 1989 (P.L. 101-218); Energy Policy Act of 2005 (EPACT 2005; P.L. 109-58); Energy Independence and Security Act of 2007 (EISA; P.L. 111-140); American Recovery and Reinvestment Act of 2009 (ARRA; P.L. 111-5)
- Annual funding: $520 million for FY2008; $472.8 million for FY2009; an estimated $1.9 billion for FY2010 from ARRA funds; an estimated $475 million for FY2011. Breakdown of additional funds appropriated from ARRA:
    - Biomass—$800 million
    - Geothermal—$400 million
    - Hydrogen/Fuel Cell—$43.4 million
    - Solar—$117.6 million
    - Wind and Hydropower—$118 million
- Scheduled termination: None
- Description: This program provides financial assistance to conduct balanced research and development efforts in the following energy technologies: solar, biomass, hydrogen, fuel cells and infrastructure, wind and hydropower, hydrogen, and geothermal. Assistance may be

used to develop and transfer renewable energy technologies to the scientific and industrial communities, states, and local governments.

- Qualified applicants: State, local, and tribal governments; colleges and universities; profit organizations; private nonprofit organizations
- Qualified technologies: Solar; biomass; hydrogen; fuel cells; wind; hydropower; geothermal
- For more information: See CFDA program number 81.087 at https://www.cfda.gov/

19. State Energy Program (SEP)

- Administered by: EERE
- Authorization: Energy Policy and Conservation Act of 1975 (EPCA; P.L. 94- 163); Energy Supply and Production Act of 1976 (ECPA; P.L. 94-385); National Energy Conservation Policy Act of 1978 (NECPA; P.L. 95-619); State Energy Efficiency Programs Improvement Act of 1990 (P.L. 101-440); Energy Policy Act of 1992 (EPACT; P.L. 102-486); Energy Conservation Reauthorization Act of 1998 (P.L. 105-388); Energy Policy Act of 2005 (EPACT 2005; P.L. 109-58); Energy Independence and Security Act of 2007 (EISA; P.L. 110-140); American Recovery and Reinvestment Act of 2009 (ARRA; P.L. 111-5)
- Annual funding: $44.1 million for FY2008; $50 million for FY2009; an additional $3.1 billion appropriated in FY2009 from ARRA; $50 million for FY2010; and $75 million requested for FY2011
- Scheduled termination: None
- Description: SEP provides grants to states to design and carry out their own renewable energy and energy efficiency programs.
- Qualified applicants: State and tribal governments, including U.S. territories
- Qualified technologies: Emerging renewable energy and energy efficiency technologies
- For more information: See EERE's State Energy Program Website at http://apps1.eere.energy and CFDA program number 81.041 at https://www.cfda.gov

20. Tribal Energy Program

- Administered by: EERE
- Authorization: Energy Policy and Conservation Act of 1975 (EPCA; P.L. 94- 163); Energy Conservation and Production Act of 1976

(ECPA; P.L. 94-385); Department of Energy Organization Act of 1977 (P.L. 95-91); Energy Tax Act of 1978 (P.L. 95-618); National Energy Conservation Policy Act of 1978 (NECPA; P.L. 95-619); Power Plant and Industrial Fuel Use Act of 1978 (P.L. 95-620); Energy Security Act (P.L. 96-294); National Appliance Energy Supply Act of 1987 (P.L. 100-12); Federal Energy Management Improvement Act of 1988 (P.L. 100-615); Energy Policy Act of 1992 (EPACT; P.L. 102-486); Energy Policy Act of 2005 (EPACT 2005; P.L. 109-58); Energy Independence and Security Act of 2007 (EISA; P.L. 110-140); American Recovery and Reinvestment Act of 2009 (ARRA; P.L. 111-5)

- Annual funding: $3.96 million for FY2006; $3.96 million for FY2007; $5.95 million for FY2008; $6 million for FY2009; $10 million for FY2010; and $10 million requested for FY2011
- Scheduled termination: None
- Description: This program promotes tribal energy sufficiency, economic growth and employment on tribal lands through the development of renewable energy and energy efficiency technologies. The program provides financial assistance, technical assistance, education and training to tribes for the evaluation and development of renewable energy resources and energy efficiency measures.
- Qualified applicant: Tribal governments
- Qualified technologies: Energy efficient technologies: clothes washers; refrigerators/freezers; water heaters; lighting; lighting and controls and sensors; chillers; furnaces; boilers; air conditioners; programmable thermostats; energy management; systems/building controls; caulking/weather-stripping; duct/air sealing; building insulation; windows; doors; siding; roofs; comprehensive measures/ whole building; and other energy efficiency improvements may be eligible. Renewable energy technologies: passive solar space heat; solar water heat; solar space heat; photovoltaics; wind; biomass; hydroelectric; geothermal electric; geothermal heat pumps
- For more information: See EERE's Tribal Energy Program website at http://apps1.eere.energy and DSIRE at http://www.dsireusa. org/ incentives/incentive.cfm?Incentive_Code=US07F&re= 1&ee=1

Other DOE Offices/Cross-Cutting Programs

21. Advanced Research Projects Energy Financial Assistance Program (ARPA-E)

- Administered by: Advanced Research Projects Agency-Energy (ARPA-E)
- Authorization: America COMPETES Act (P.L. 110-69), Section 5012
- Annual funding: $15 million for FY2009; an additional $388.9 million in FY2009 from ARRA; $0 for FY2010; and $300 million requested for FY2011
- Scheduled termination: After ARPA-E has been in operation for four years, the Secretary of Energy shall offer to enter into a contract with the National Academy of Sciences under which the National Academy shall conduct an evaluation of how well ARPA-E is achieving the goals and mission of ARPA-E. The evaluation shall include the recommendation of the National Academy of Sciences on whether ARPA-E should be continued or terminated.
- Description: This program will fund organizations that have proposed sophisticated energy technology R&D projects that (1) translate scientific discoveries and cutting-edge inventions into technological innovations and (2) accelerate transformational technological advances in areas that industry by itself is not likely to undertake because of high technical or financial risk. Transformational energy technologies are those that have the potential to create new paradigms in how energy is produced, transmitted, used, or stored.
- Qualified applicants: ARPA-E welcomes submissions from any type of capable technology research and development entity. This includes, but is not limited to for-profit entities, academic institutions, research foundations, not-for-profit entities, collaborations, and consortia. The lead organization that will enter into the agreement with ARPA-E must be a U.S. entity.
- Qualified technologies: Transformational energy technologies
- For more information: See ARPA-E Frequently Asked Questions (FAQ) Website at http://arpa-e.energy and CFDA program number 81.135 at https://www.cfda.gov

22. Electricity Delivery and Energy Reliability, Research, Development and Analysis Grant Program (Office of Electricity Delivery and Energy Reliability)

- Administered by: Office of Electricity Delivery and Energy Reliability (OE)
- Authorization: Department of Energy Organization Act of 1977 (P.L. 95-91); Energy Security Act of 1980 (P.L. 96-294); National Superconductivity and Competitiveness Act of 1988 (P.L. 100-697); Energy Policy Act of 1992 (EPACT; P.L. 102-486); Energy Policy Act of 2005 (EPACT 2005; P.L. 109- 58); Energy Independence and Security Act of 2007 (EISA; P.L. 110-140); American Recovery and Reinvestment Act of 2009 (ARRA; P.L. 111-5)
- Annual funding: \$82.8 million for FY2008; \$83.1 million for FY2009; an additional \$4.5 billion was appropriated to the Office of Electricity Delivery and Energy Reliability in FY2009 from ARRA. Approximately \$4 billion of that total was used to implement smart grid programs authorized by EISA and accelerate the deployment of smart grid technologies across the transmission and distributions[1]; \$125 million for FY2010; and \$144.3 million requested for FY2011
- Scheduled termination: None
- Description: This grant program aims to develop cost-effective technology that enhances the reliability, efficiency, and resiliency of the electric grid.
- Qualified applicants: State, local, and tribal governments; universities; profit organizations; private nonprofit organizations; research organizations
- Qualified technologies: Specific technologies not listed
- For more information: See CFDA Program number 81-122 at https://www.cfda.gov

23. Federal Energy Management Program (FEMP)

- Administered by: EERE
- Authorization: Energy Policy and Conservation Act of 1975 (EPCA; P.L. 94- 163); Energy Conservation and Production Act of 1976 (ECPA; P.L. 94-385); Department of Energy Organization Act (P.L. 95-91); National Energy Conservation Policy Act of 1978 (NECPA; P.L. 95-619); Federal Energy Management Improvement Act of 1988 (P.L. 100-615); Energy Policy Act of 1992 (EPACT; P.L. 102-486);

Energy Policy Act of 2005 (EPACT 2005; P.L. 109-58); Energy Independence and Security Act of 2007 (EISA; P.L. 110-140)

- Annual funding: $19 million for FY2006; $19.5 million for FY2007; $19.8 million for FY2008; $22 million for FY2009; an additional $22.4 million in FY2009 from ARRA; $32 million for FY2010; and $42.3 million requested for FY2011
- Scheduled termination: None
- Description: FEMP assists federal agencies in developing and implementing energy efficient and renewable energy resources to meet energy management regulations and goals.
- Qualified applicants: Federal agencies
- Qualified technologies: Energy efficient technologies; solar; wind; incremental hydro; ocean; biomass; geothermal
- For more information: See EERE's Federal Energy Management Program Website at http://www1.eere.energy

24. Financial Assistance Program (Office of Science)

- Administered by: Office of Science
- Authorization: Atomic Energy Act of 1954 (P.L. 83-703), Section 31; Energy Reorganization Act of 1974 (P.L. 93-438), Title I, Section 107; Federal Nonnuclear Energy Research and Development Act of 1974 (P.L. 93-577); American Recovery and Reinvestment Act of 2009 (ARRA; P.L. 111-5)
- Annual funding: $974 million for FY2008; $1.4 billion for FY2009; an estimated $1.3 billion for FY2010; and an estimated $1.3 billion for FY2011
- Scheduled termination: None
- Description: This program provides financial support for fundamental research in the basic sciences and advanced technology concepts and assessments in fields related to energy.
- Qualified applicants: State, local, and tribal governments; colleges and universities; profit commercial organizations; private nonprofit organizations; public nonprofit organizations; small businesses
- Qualified technologies: Specific advanced technologies not listed
- For more information: See CFDA Program number 81.049 at https://www.cfda.gov

25. Loan Guarantee Program (Office of the Chief Financial Officer)

- Administered by: Office of the Chief Financial Officer
- Authorization: Energy Policy Act of 2005 (EPACT 2005; P.L. 109-58), Title XVII; American Recovery and Reinvestment Act of 2009 (ARRA; P.L. 111-5); 42 USC 16511 et seq.
- Annual funding:
- Section 1703 Innovative Technology Loan Guarantee Program (permanent): $4.5 million for FY2008; $0 for FY2009; $0 for FY2010; and a $500 million request for FY2011
- Section 1705 Temporary Loan Guarantee Program: $0 for FY2008; $6 billion was appropriated for FY2009. However, $2 billion of that funding has since been transferred to the "cash for clunkers" automobile trade-in program by P.L. 111- 47.[2] An additional $1.5 billion was rescinded for the Education Jobs and Medicaid Assistance Act, P.L. 111-226 (section 308), leaving a total of $2.5 billion remaining from the FY2009 appropriations; $0 for FY2010; and $0 requested for FY2011.
- Scheduled termination: None for the permanent (Section 1703) loan guarantee program. Projects authorized by the temporary loan guarantee (Section 1705) must begin construction no later than September 30, 2011.
- Description: This program provides federal loan guarantees to encourage early commercial use in the United States of new or significantly improved technologies in energy projects that (1) avoid, reduce, or sequester air pollutants or anthropogenic emissions of greenhouse gases; and (2) employ new or significantly improved technologies as compared to commercial technologies in service in the United States at the time the guarantee is issued. Temporary loan guarantees can also be made under Section 1705 for rapid deployment of certain renewable and electric transmission projects.
- Qualified applicants: State, local, and tribal governments; universities; profit organizations; and public nonprofit organizations. No federal entity may apply.
- Qualified technologies: Solar thermal electric; solar thermal process heat; photovoltaics; wind; hydroelectric; renewable transportation fuels; geothermal electric; fuel cells; manufacturing facilities; daylighting; tidal energy; wave energy; ocean thermal; biodiesel.
- For more information: See CFDA Program number 81.126 at https://www.cfda.gov; DSIRE at http://www.dsireusa.org/incentives/

incentive.cfm?Incentive_Code=US48F&re=1&ee=1;and DOE's Loan Guarantee Program website at http://www.lgprogram. energy. gov/

26. Small Business Innovation Research Program (SBIR)/Small Business Technology Transfer Program (STTR)

- Administered by: Office of Science
- Authorization: Small Business Innovation Development Act of 1982 (P.L. 97- 219); Small Business Research and Development Act of 1992 (P.L. 102-564); Consolidated Appropriations Act, 2001 (P.L. 106-554), Title I, Small Business Innovation Research Program Reauthorization Act of 2000; Small Business Technology Transfer Program Reauthorization Act of 2001 (P.L. 107-50)
- Annual funding: $116 million for FY2007 ($36 million for Phase I, $80 million for Phase II); $116 million for FY2008 ($36 million for Phase I, $80 million for Phase II); $116 million for FY2009 ($36 million for Phase I, $80 million for Phase II); approximately $94 million for FY2010 ($36 million for Phase I, $58 million for Phase II); approximately $38 million anticipated appropriations for Phase I funding opportunities in FY2011; Anticipated appropriations for Phase II funding opportunities for FY2011 is not currently available
- Scheduled termination: The SBIR and STTR Acts expired September 30, 2009. SBIR or STTR grants may not be awarded until the programs are extended or renewed by public law
- Description: Small Business Innovation Research (SBIR) and Small Business Technology Transfers (STTR) are U.S. government programs in which federal agencies with large research and development (R&D) budgets set aside a small fraction of their funding for competitions among small businesses only. Small businesses that win awards in these programs keep the rights to any technology developed and are encouraged to commercialize the technology.
- Qualified applicants: Small businesses
- Qualified technologies: Research areas include energy production (fossil, nuclear, renewable, and fusion energy); energy use (in buildings, vehicles, and industry), fundamental energy sciences (materials, life, environmental, and computational sciences, and nuclear and high energy physics); environmental management; and nuclear nonproliferation

- For more information: See DOE's Small Business Innovation Research (SBIR) and Small Business Technology Transfers (STTR) website at http://www.sc.doe.gov/sbir/FAQ.html; and CFDA Program number 10.212 (SBIR) at https://www.cfda.gov

II. U.S Department of the Treasury

Homeowner

1. Residential Energy Efficiency Tax Credit

- Administered by: Internal Revenue Service
- Authority: Energy Policy Act of 2005 (EPACT 2005; P.L. 109-58); Energy Improvement and Extension Act of 2008 (P.L. 110-343), Division B, Section 302; American Recovery and Reinvestment Act of 2009 (ARRA; 111-5), Division B, Section 1121; 26 USC 25C
- Scheduled Termination: December 31, 2011 ($500 credit limit); the $1,500 credit limit expired on December 31, 2010
- Description: The credit applies to energy efficiency improvements in the building envelope of existing homes and for the purchase of high-efficiency heating, cooling and water-heating equipment. Efficiency improvements or equipment must serve a dwelling in the United States that is owned and used by the taxpayer as a primary residence. The maximum amount of homeowner credit for all improvements combined for 2011 has been reduced from the $1,500 for purchases in 2009 and 2010 to $500 total for equipment purchased between 2006 and 2011.
- Qualified applicant: Residential
- Qualifying technologies: Water heaters; furnace; boilers; heat pumps; air conditioners; building insulation; windows; doors; roofs; circulating fans used in a qualifying furnace; biomass and stoves that use qualified biomass fuel
- For more information: See the Internal Revenue Service website at http://www.irs.gov/formspubs/article/0,,id=207332,00.html

2. Residential Renewable Energy Tax Credit

- Administered by: Internal Revenue Service
- Authority: Energy Policy Act of 2005 (EPACT 2005; 109-58); Energy Improvement and Extension Act of 2008 (P.L. 110-343); American

Recovery and Reinvestment Act of 2009 (ARRA; P.L. 111-5); 26 USC 25D (amended)

- Scheduled Termination: December 31, 2016
- Description: A taxpayer may claim a credit of 30% of qualified expenditures for a system that serves a dwelling unit located in the United States and used as a residence by the taxpayer.
- Qualified applicant: Residential
- Qualifying technologies: Solar water heat; photovoltaics; wind; fuel cells; geothermal heat pumps; other solar electric technologies
- For more information: See IRS Form 5695 & Instructions: Residential Energy Credits at http://www.irs.gov/pub/irs-pdf/f5695.pdf

Business

3. Business Energy Investment Tax Credit

- Administered by: Internal Revenue Service
- Authority: Energy Improvement and Extension Act of 2008 (EISA; P.L. 110- 343), Division B; American Recovery and Reinvestment Act of 2009 (ARRA; P.L. 111-5), Division B, Section 1103
- Scheduled termination: December 31, 2016. Geothermal property, with the exception of geothermal heat pumps, has no stated expiration date. The credit for solar energy property returns to 10% after December 31, 2016.
- Description: Credit is 30% for solar, fuel cells and small wind; 10% for
  geothermal, microturbines and CHP (Combined Heat and Power)
- Qualified Applicants: Commercial; industrial; utilities
- Qualified Technologies: Solar water heat; solar space heat; solar thermal electric; solar thermal process heat; photovoltaics; wind; biomass; geothermal electric; fuel cells; geothermal heat pumps; CHP/Cogeneration; solar hybrid lighting; direct-use geothermal; microturbines
- For more information see the DSIRE website at http://www. dsireusa. org/incentives/incentive. cfm? Incentive_Code= US02F&re=1&ee=0

4. Energy Efficient Commercial Buildings Tax Deduction

- Administered by: Internal Revenue Service
- Authority: Energy Policy Act of 2005 (EPACT 2005; P.L. 109-58), Title XIII, Subtitle C, Section 1331(a); Tax Relief and Health Care Act of 2006 (P.L. 109- 432), Division A, Title II, Section 204; Energy Improvement and Extension Act of 2008 (P.L. 110-343), Division B, Title III, Section 303; 26 USC § 179D Scheduled termination: December 31, 2013
- Description: A tax deduction of $1.80 per square foot is available to owners of new or existing buildings who install (1) interior lighting; (2) building envelope, or (3) heating, cooling, ventilation, or hot water systems that reduce the building's total energy and power cost by 50% or more in comparison to a building meeting minimum requirements set by ASHRAE Standard 90.1-2001. Energy savings must be calculated using qualified computer software approved by the IRS.
- Qualified applicants: Commercial; builder/developer; state government; federal government (deductions associated with government buildings are transferred to the designer)
- Qualified technologies: Equipment insulation; water heaters; lighting; lighting controls/sensors; chillers; furnaces; boilers; heat pumps; air conditioners; caulking/weather-stripping; duct/air sealing; building insulation; windows; doors; siding; roofs; comprehensive measures/ whole building
- For more information: See the Energy Star website at http://www.energystar.gov/index.cfm?c=tax_credits.tx_comm_buildings

5. Energy-Efficient New Homes Tax Credit for Home Builders

- Administered by: Internal Revenue Service
- Authority: Tax Technical Corrections Act of 2007 (P.L. 110-172), Section 11(a)(7); Energy Improvement and Extension Act (P.L. 110-343), Division B, Title III, Section 304; 26 USC § 45L 8/8/2005 (amended 2008); P.L. 111-312 (12/31/2011 extension)
- Scheduled termination: December 31, 2011
- Description: This program provides tax credits of up to $2,000 for builders of all new energy-efficient homes, including manufactured homes constructed in accordance with the Federal Manufactured Homes Construction and Safety Standards. This credit was created by the Energy Policy Act of 2005 for homes constructed in 2006 and 2007. It was renewed for homes constructed in 2008 and 2009, but

then it expired and then was not active in 2010. The Tax Relief, Unemployment Insurance Reauthorization, and Job Creation Act of 2010 (P.L. 111-312) retroactively reinstated this credit for homes acquired after December 31, 2009 and before January 1, 2012.

- Qualified applicant: Builder/developer
- Qualified technologies: Comprehensive measures/whole building
- For more information: See the IRS website at http://www.irs.gov/ pub/irs-pdf/ f8908.pdf; http://www.irs.gov/pub/irs-drop/n-06-27.pdf; http://www.irs.gov/ pub/irs-drop/n-06-28.pdf

6. Renewable Energy Grants (1603 Program)

- Administered by U.S. Department of the Treasury
- Authority: Sec. 707 of H.R. 4853, Tax Relief, Unemployment Insurance Reauthorization, and Job Creation Act of 2010; H.R. 1, American Recovery and Reinvestment Act of 2010 (ARRA; P.L. 111-5) Division B, Sec 1104 & 1603, U.S. Department of Treasury: Grant Program Guidance (amended)
- Scheduled Termination: Construction must begin by December 31, 2011. Applications must be submitted before October 1, 2012.
- Description: Renewable energy grant program. To be eligible, a property must be placed in service in 2009, 2010, or 2011 or placed in service after 2011 but only if construction of the property began during 2009, 2010, or 2011. The credit termination date remains unchanged. Applications must be submitted before October 1, 2012.
- The American Recovery and Reinvestment Act of 2009 (ARRA; P.L. 111-5) allows taxpayers eligible for the federal business energy investment tax credit (ITC) to take this credit or to receive a grant from the U.S. Treasury Department instead of taking the business ITC for new installations. The new law also allows taxpayers eligible for the renewable electricity production tax credit (PTC) to receive a grant from the U.S. Treasury Department instead of taking the PTC for new installations. (It does not allow taxpayers eligible for the residential renewable energy tax credit to receive a grant instead of taking this credit.) Taxpayers may not use more than one of these incentives. Tax credits allowed under the ITC with respect to progress expenditures on eligible energy property will be recaptured if the project receives a grant. The grant is not included in the gross income of the taxpayer. This grant cannot be taken for systems where construction began after December 31, 2011.

- Qualified applicants: Commercial, Industrial, Agricultural
- Qualified technologies: Solar Water Heat, Solar Space Heat, Solar Thermal Electric, Solar Thermal Process Heat, Photovoltaics, Landfill Gas, Wind, Biomass, Hydroelectric, Geothermal Electric, Fuel Cells, Geothermal Heat Pumps, Municipal Solid Waste, CHP/Cogeneration, Solar Hybrid Lighting, Hydrokinetic, Anaerobic Digestion, Tidal Energy, Wave Energy, Ocean Thermal, Microturbines
- More information: See the Treasury's 1603 website at http://www.treasury.gov/initiatives/recovery/Pages/1603.aspx;program guidance http://www.treasury.gov/initiatives/recovery/Documents/guidance.pdf

Industry

7. Energy Efficient Appliance Tax Credit for Manufacturers

- Administered by: Internal Revenue Service
- Authority: Energy Policy Act of 2005 (EPACT 2005; P.L. 109-58), Title XIII, Subtitle C, Section 1334(a); Energy Improvement and Extension Act of 2008 (P.L. 110-343), Division B, Section 305; 26 USC§45M subsequently amended; H.R. 4853, Tax Relief, Unemployment Insurance Reauthorization, and Job Creation Act of 2010
- Scheduled termination: December 31, 2011
- Description: A tax credit for new appliances that meet Energy Star 2007 requirements. Each manufacturer is limited to a total of $25 million in 2011 for all credits under this provision.
- Qualified applicants: Industrial; appliance manufacturers
- Qualified technologies: Clothes washers; dishwashers; refrigerators
- For more information: See the IRS website at http://www.irs.gov/businesses corporations/article/0,,id=208024,00. html; IRS form 8909 at http://www.irs.gov/pub/irs-pdf/f8909.pdf

8. Qualifying Advanced Energy Manufacturing Investment Tax Credit (48C)

- Administered by: Internal Revenue Service
- Authority: American Recovery and Reinvestment Act of 2009 (ARRA; P.L. 111-5), Division B, Section 1302; 26 USC 48C
- Scheduled termination: The applications were due to DOE by September 16, 2009, with final applications due to DOE October 16, 2009. This incentive is no longer available; an act of Congress is

required to renew this tax incentive. As of February 2011, this has not yet occurred.

- Description: The U.S. Treasury Department, in consultation with the U.S. Department of Energy (DOE), is no longer accepting applications for this tax credit. The applications were due to DOE by September 16, 2009, with final applications due to DOE October 16, 2009. Only applicants accepted and ranked by the DOE were allowed to submit final applications to the Internal Revenue Service (IRS) by December 16, 2009. Approved projects were announced in January 2010.
- This tax credit was designed to encourage a U.S.-based renewable energy manufacturing sector. Projects receiving awards are eligible for a tax credit of 30% of the qualified investment required for an advanced energy project. Qualified applicants: Commercial; industrial; manufacturing.
- Qualified applicants: Commercial, industrial, manufacturing
- Qualifying technologies: Lighting; lighting controls/sensors; energy conservation technologies: smart grid; solar water heat; solar thermal electric; photovoltaics; wind; geothermal electric; fuel cells; geothermal heat pumps; batteries and energy storage; advanced transmission technologies that support renewable energy generation; renewable fuels; fuel cells using renewable fuels; microturbines
- For more information: See the U.S. Department of Energy, http://www.energy and http://www.energy.gov/ recovery and http://www. whitehouse.gov/the-press-office/president billion-new-clean-tech-manufacturing-jobs

9. Renewable Electricity Production Tax Credit (PTC)

- Administered by: Internal Revenue Service
- Authority: Energy Policy Act of 1992 (EPACT; P.L. 102-486); Energy Policy Act of 2005 (EPACT 2005; P.L. 109-58), Title XIII, Section 1301; Tax Relief and Health Care Act of 2006 (P.L. 109-432), Division A, Section 201; Energy Improvement and Extension Act of 2008 (P.L. 110-343); American Recovery and Reinvestment Act of 2009 (ARRA; P.L. 111-5), Division B, Section 1101 and 1102, 26 USC 45 (amended)
- Scheduled termination: The duration of the credit is generally 10 years after the date the facility is placed in service, but there are two exceptions: Open-loop biomass, geothermal, small irrigation hydro,

landfill gas and municipal solid waste combustion facilities placed into service after October 22, 2004, and before enactment of the Energy Policy Act of 2005, on August 8, 2005, are only eligible for the credit for a five-year period. Open-loop biomass facilities placed in service before October 22, 2004, are eligible for a five-year period beginning January 1, 2005

- Description: The federal renewable electricity Production Tax Credit (PTC) is a per-kilowatt-hour tax credit for electricity generated by qualified energy resources and sold by the taxpayer to an unrelated person during the taxable year. The American Recovery and Reinvestment Act of 2009 (ARRA; P.L. 111- 5) allows taxpayers eligible for the federal renewable electricity production tax credit (PTC) to take the federal business energy investment tax credit (ITC) or to receive a grant from the U.S. Treasury Department instead of taking the PTC for new installations. The grant is only available to systems where construction begins prior to December 31, 2011. The new law also allows taxpayers eligible for the business ITC to receive a grant from the U.S. Treasury Department instead of taking the business ITC for new installations. The Treasury Department issued Notice 2009-52 in June 2009, giving limited guidance on how to take the federal business energy investment tax credit instead of the federal renewable electricity production tax credit.
- Qualified applicants: Commercial; industrial
- Qualifying technologies: Landfill gas; wind; biomass; hydroelectric; geothermal electric; municipal solid waste; hydrokinetic power (i.e., flowing water); anaerobic digestion; small hydroelectric; tidal energy; wave energy; ocean thermal.
- For more information: See the IRS website at http://www.irs.gov/ pub/irs-pdf/ f8835.pdf

10. Residential Energy Conservation Subsidy Exclusion (Corporate)

- Administered by: Internal Revenue Service
- Authority: Energy Policy Act of 1992 (EPACT; P.L. 102-486), Section 1912; Small Job Protection Act of 1996 (P.L. 104-188), Section 1617; 26 USC 136 (amended)
- Scheduled termination: None
- Description: Energy conservation subsidies provided by public utilities, either directly or indirectly, are nontaxable: "Gross income shall not include the value of any subsidy provided (directly or

indirectly) by a public utility to a customer for the purchase or installation of any energy conservation measure."

- Qualified applicants: Residential; multi-family residential
- Qualifying technologies: Solar water heat; solar space heat; photovoltaics; other energy efficiency technologies not identified.
- For more information: See the IRS website at http://www.irs.gov/publications/ p525/index.html

11. Residential Energy Conservation Subsidy Exclusion (Personal)

- Administered by: Internal Revenue Service
- Authority: Energy Policy Act of 1992 (EPACT; P.L. 102-486); Small Job Protection Act of 1996 (P.L. 104-188), Section 1617; 26 USC 136 (amended)
- Scheduled termination: None
- Description: Energy conservation subsidies provided by public utilities, either directly or indirectly, are nontaxable: "Gross income shall not include the value of any subsidy provided (directly or indirectly) by a public utility to a customer for the purchase or installation of any energy conservation measure."
- Qualified applicant: Residential; multi-family residential
- Qualifying technologies: Specific efficiency technologies not identified; eligible renewables: Solar water heat; solar space heat; photovoltaics
- For more information: See the IRS website at http://www.irs.gov/publications/ p525/index.html

State, Local and Tribal Governments

12. Clean Renewable Energy Bonds (CREBs)

- Administered by: Internal Revenue Service
- Authority: Established by the Energy Policy Act of 2005 (EPACT 2005; P.L. 109-58); Tax Relief and Health Care Act of 2006 (P.L. 109-432); Energy Improvement and Extension Act of 2008 (P.L. 110-343), Division B; American Recovery and Reinvestment Act of 2009 (ARRA; P.L. 111-5); 26 USC 54 (old CREBs), 26 USC 54A (new CREBs); 26 USC 54C (new CREBs); IRS Notice 2009-33; IRS Announcement 2010-54
- Annual funding: EPACT originally allocated $800 million of tax credit bonds to be issued between January 1, 2006, and December 31,

2007. Following the enactment of the federal Tax Relief and Health Care Act of 2006, the IRS made an additional $400 million in CREBs financing available for 2008 through Notice 2007-26. In November 2006, the IRS announced that the original $800 million allocation had been reserved for a total of 610 projects. The additional $400 million (plus surrendered volume from the previous allocation) was allocated to 312 projects in February 2008. Of the $1.2 billion total of tax-credit bond volume cap allocated to fund renewable-energy projects, state and local government borrowers were limited to $750 million of the volume cap, with the rest reserved for qualified municipal or cooperative electric companies. The Energy Improvement and Extension Act of 2008 (Div. A, Sec. 107) allocated $800 million for new CREBs. In February 2009, the American Recovery and Reinvestment Act of 2009 (Div. B, Sec. 1111) allocated an additional $1.6 billion to expand the total new CREBs allocation to $2.4 billion

- Scheduled termination: November 1, 2010 (New CREBs Electric Cooperatives Solicitation deadline, expired)
- Description: The IRS is not currently accepting applications for New CREB bond volume. CREBs are used to finance renewable energy projects. CREBs are issued, theoretically, with a 0% interest rate. The borrower pays back only the principal of the bond and the bondholder receives federal tax credits in lieu of the traditional bond interest.
- Qualified applicants: State, local, and tribal governments; municipal utility; rural electric cooperative
- Qualified technologies: Solar thermal electric; photovoltaics; landfill gas; wind; biomass; hydroelectric; geothermal electric; municipal solid waste; hydrokinetic power; anaerobic digestion; tidal energy; wave energy; ocean thermal
- For more information: See the DSIRE website at http://www. dsireusa.org/ incentives/incentive.cfm? Incentive_Code= US45F&re= 1&ee=1; and the Internal Revenue Service Website at http://www.irs. gov/irb/2007-14_IRB/ar17.html and http://www. irs. gov/pub/irs-drop/n-09-33.pdf

13. Qualified Energy Conservation Bonds (QECBs)

- Administered by: Internal Revenue Service
- Authority: Energy Improvement and Extension Act of 2008 (P.L. 110-343), Division B, Section 301; American Recovery and Reinvestment

Act of 2009, (ARRA; P.L. 111-5), Division B, Title I, Subtitle F, Part III, Section 1521(b)(1), (2), and Part IV, Section 1531(c)(2), and Part V, Section 1541(b)(2); 26 USC 54A ; 26 USC 54D; IRS Notice 2009-29; IRS Notice 2010-35

- Scheduled termination: None
- Description: QECBs may be used by state, local and tribal governments to finance certain types of energy projects. These bonds are issued, theoretically, with a 0% interest rate. The borrower pays back only the principal of the bond and the bondholder receives federal tax credits in lieu of the traditional bond interest. The tax credit may be taken quarterly to offset the tax liability of the bondholder. The tax credit rate is set daily by the U.S. Treasury Department; however, energy conservation bondholders will receive only 70% of the full rate set by the Treasury Department under 26 USC 54A. Credits exceeding a bondholder's tax liability may be carried forward to the succeeding tax year, but cannot be refunded. Energy conservation bonds differ from traditional tax-exempt bonds in that the tax credits issued through the program are treated as taxable income for the bondholder. The original limit on the volume of energy conservation tax credit bonds to be issued by state and local governments was $800 million. However, The American Recovery and Reinvestment Act of 2009, enacted in February 2009, expanded the allowable bond volume to $3.2 billion.
- Qualified applicants: State, local, and tribal governments
- Qualified technologies: Solar thermal electric; photovoltaics; landfill gas; wind; biomass; hydroelectric; geothermal electric; municipal solid waste; hydrokinetic power; anaerobic digestion; tidal energy; wave energy; ocean thermal
- For more information: See the DSIRE website at http://www.dsireusa.org/ incentives/ incentive.cfm? Incentive_Code= US51F&re=1&ee=1

Cross-Cutting

14. Modified Accelerated Cost-Recovery System (MACRS) + Bonus Depreciation (2008-2012)

- Administered by: Internal Revenue Service
- Authority: Economic Recovery Tax Act of 1981 (P.L. 97-34); Economic Stimulus Act of 2008 (P.L. 110-185); American Recovery and Reinvestment Act of 2009 (ARRA; P.L. 111-5), Division B, Title I, Subtitle C, Part I, Section 1201(a)(1)-(2)(D), (3)(A), (b)(1); Tax Relief, Unemployment Insurance Reauthorization, and Job Creation Act of 2010 (P.L. 111-312); 26 USC 168; 26 USC 48
- Scheduled termination: The five-year schedule for most types of solar, geothermal, and wind property has been in place since 1986. December 31, 2011 (100% bonus depreciation); December 31, 2012 (50% bonus depreciation)
- Description: Under MACRS, businesses may recover investments in certain property through depreciation deductions. The MACRS establishes a set of class lives for various types of property, ranging from three to 50 years, over which the property may be depreciated. A number of renewable energy technologies are classified as five-year property (26 USC 168(e)(3)(B)(vi)) under the MACRS, which refers to 26 USC 48(a)(3)(A), often known as the Energy Investment Tax Credit or ITC to define eligible property.
- Qualified applicants: Commercial; industrial
- Qualified technologies: Solar water heat; solar space heat; solar thermal electric; solar thermal process heat; photovoltaics; landfill gas; wind; biomass; renewable transportation fuels; geothermal electric; fuel cells; geothermal heat pumps; municipal solid waste; CHP/cogeneration; solar hybrid lighting; direct use geothermal; anaerobic digestion; microturbines
- For more information: See the IRS website at http://www.irs. gov/ publications/ p946/ch04.html

15. Alternative Motor Vehicle Credit

- Administered by: Internal Revenue Service
- Authority: Energy Policy Act of 2005 (EPACT 2005; P.L. 109-58); Energy Improvement and Extension Act of 2008 (P.L. 110-343), Division B, Section 205; American Recovery and Reinvestment Act of 2009 (ARRA; P.L. 111-5), Division B, Sections 1141-1144

- Scheduled Termination: The IRS advises taxpayers that provisions of the Alternative Motor Vehicle Credit including the Qualified Hybrid Motor Vehicle Credit, Qualified Alternative Fuel Motor Vehicle Credit (QAFMV), and Advanced Lean Burn Technology Motor Vehicle Credit expired as of December 31, 2010; the Qualified Plug-In Electric Motor Vehicle Conversion Credit expires on December 31, 2011; and the Qualified Plug-In Electric Drive Motor Vehicle Credit and the Fuel Cell Motor Vehicle Credit both expire on December 31, 2014.
- Description: A tax credit is subtracted directly from the total amount of federal tax owed, thus reducing or even eliminating the taxpayer's tax obligation. The tax credit for hybrid vehicles applies to vehicles purchased or placed in service on or after January 1, 2006. Purchasers of advanced lean burn technology motor vehicles may claim a credit of $1,300 per vehicle.
- Qualified applicant: Taxpayers
- Qualifying technologies: Hybrid gasoline-electric; diesel; battery-electric; alternative fuel and fuel cell vehicles; advanced lean-burn technology vehicles; plug-in hybrid electric vehicles
- For more information: See the IRS website for the Alternative Motor Vehicle Credit at http://www.irs.gov/businesses/corporations/ article/ 0,,id= 202341,00.html

III. DEPARTMENT OF AGRICULTURE

1. Assistance to High Energy Cost Rural Communities Program

- Administered by: Rural Development
- Authorization: Rural Electrification Act of 1936 (P.L. 74-605); Grain Standards and Warehouse Improvement Act of 2000 (P.L. 106-472)
- Annual funding: $18 million for FY2009; $18 million for FY2010; FY2011 budget request was $0
- Scheduled termination: None
- Description. This program provides financial assistance to rural communities with extremely high energy costs (exceeding 275% of the national average)

- Qualified applicants: State, local, and tribal governments (including U.S. territories); for-profit businesses; non-profit businesses; cooperatives; individuals
- Qualified technologies: Not specifically identified
- For more information: See CFDA program number 10.859 at https://www.cfda.gov and the USDA program website at http://www. usda. gov/rus/electric/hecgp/index.htm

2. Bioenergy Program for Advanced Biofuels

- Administered by: Rural Development
- Authorization: P.L. 110-234, Food, Conservation, and Energy Act of 2008, Title IX, section 9001, subsection 9005
- Annual Funding:
- Mandatory Farm Bill authorization: $55 million for FY2009; $55 million for FY2010; $85 million for FY2011 Discretionary: $25 million authorized for FY2009-FY2011; $0 appropriated for FY2009-FY2011
- Scheduled Termination: Mandatory funding authorized through FY2012
- Description: To support and ensure an expanding production of advanced biofuels by providing payments to eligible advanced biofuel producers.
- Qualified applicants: Eligible advanced biofuels producers
- Qualified technologies: Payments will be made to eligible advanced biofuel producers for the production of fuel derived from renewable biomass, other than corn kernel starch, to include biofuel derived from cellulose, hemicellulose, or lignin; biofuel derived from sugar and starch (other than Ethanol derived from corn kernel starch); biofuel derived from waste material, including crop residue, other vegetative waste material, animal waste, food waste and yard waste; diesel-equivalent fuel derived from renewable biomass, including vegetable oil and animal fat; biogas (including landfill gas and sewage waste treatment gas) produced through the conversion of organic matter from renewable biomass; butanol or other alcohols produced through the conversion of organic matter from renewable biomass; and other fuel derived from cellulosic biomass

- For more Information: See CFDA program number 10.867 at https://www.cfda.gov and the USDA Program website http://www.rurdev.usda.gov/rbs/busp/9005Biofuels.htm

3. Biomass Crop Assistance Program (BCAP; Sec. 9011)

- Administered by: Farm Services Agency (FSA)
- Authorization: Title IX of the Farm Security and Rural Investment Act of 2002 (FSRIA; P.L. 107-171) is amended by Title IX, section 9001 of the Food, Conservation, and Energy Act of 2008 (P.L. 110-246), creating new section 9011 under FSIRA
- Annual funding: Original mandatory funding authorization for FY2009-FY2012 authorizes "such sums as necessary." The Supplemental Appropriations Act of 2010 (P.L. 111-212) limits mandatory spending on BCAP by allowing no more than $552 million in FY2010 and $432 million in FY2011. Funding in 2012 remains at "such sums as necessary."
- Scheduled termination: Funding authorized through FY2012
- Description: BCAP provides assistance to support the production of eligible biomass crops on land within approved BCAP project areas. In exchange for growing eligible crops, the FSA will provide annual payments through 10- to 15-year contracts. Under these contracts up to 75% of establishment costs may also be provided. FSA will also provide matching payments to eligible material owners at a rate of $1 for each $1 per dry ton paid by a qualified biomass conversion facility. Payments may not exceed $45 per ton for a two-year period and matching payments are available for no more than two years per participant.
- Qualified applicants: Eligible biomass material owners and eligible biomass producers
- Qualified technologies: Eligible material for a matching payment is renewable biomass, as defined by the 2008 farm bill, with several important exclusions including harvested grains, fiber or other commodities eligible to receive payments under the Commodity
- Title (Title I) of the 2008 farm bill (the residues of these commodities, however, are eligible and may qualify for payment); animal waste and animal waste by-products including fats, oils, greases and manure; food waste and yard waste; and algae. Eligible crops include

renewable biomass, with the exception of crops eligible to receive a payment under Title I of the 2008 farm bill and plants that are invasive or noxious, or have the potential to become invasive or noxious.

- For more Information: See the USDA website at http://www.fsa.usda.gov/FSA/webapp? area=home&subject=ener& topic=bcap and http://www.apfo.usda.gov/Internet/FSA_File/bcap2010.pdf;

4. Biorefinery Assistance Program (Sec. 9003)

- Administered by: Rural Development
- Authorization: Title IX of the Farm Security and Rural Investment Act of 2002 (FSRIA, P.L. 107-171) is amended by Title IX, section 9001 of the Food, Conservation and Energy Axt of 2008 (P.L. 110-246) creating new section 9003 under FSIRA
- Annual Funding:
- Mandatory authorization: $75 million for FY2009; $245 million for FY2010. Mandatory funding to remain available until expended.
- Discretionary authorization: $150 million authorized annually for FY2009-FY2012; $0 discretionary funding appropriated FY2009-FY2011
- Scheduled Termination: Funding authorized through FY2012
- Description: The purpose is to assist in the development of new and emerging technologies for the development of advanced biofuels, so as to increase the energy independence of the United States; promote resource conservation, public health, and the environment; diversify markets for agricultural and forestry products and agriculture waste material; and create jobs and enhance the economic development of the rural economy. Loan guarantees are made to fund the development, construction, and retrofitting of commercial-scale biorefineries using eligible technology. The maximum loan guarantee is $250 million.
- Qualified applicants: Individuals, tribal entities, state government entities, local government entities, corporations, farm cooperatives, farmer cooperative organizations, associations of agricultural producers, national laboratories, institutions of higher education, rural electric cooperatives, public power entities, and consortia of any of the previous entities

- Qualified technologies: Technologies being adopted in a viable commercial-scale operation of a biorefinery that produces an advanced biofuel; and technologies that have been demonstrated to have technical and economic potential for commercial application in a biorefinery that produces an advanced biofuel
- For more Information: See the USDA website at http://www.rurdev.usda.gov/rbs/busp/ baplg9003.htm; CFDA program number 10.865 at https://www.cfda.gov;

5. Community Wood Energy Program

- Administered by: Forest Service
- Authorization: Title IX of the Farm Security and Rural Investment Act of 2002 (FSRIA, P.L. 107-171 is amended by Title IX section 9001 of the Food, Conservation, and Energy Act of 2008 (P.L. 110-246), creating new section 9013 under FSIRA
- Annual funding: Discretionary authority: $5 million authorized annually for FY2009-FY2012; $0 appropriated for FY2009-FY2011. The FY2012 President's budget proposed to fund the program using funds from the Hazardous Fuels program within the Forest Service.Scheduled termination: Authorized through FY2012
- Description: Grants awarded for systems smaller than 5 million Btu per hour for heating (or 2 megawatts) for electric power production as directed by statute. At least a 50% match is required from Non-Federal funds for grants. Technical assistance will be based on previous work and commitment to future work demonstrated by the applicant. The program is authorized $5 million annually when funded. Grant awards are limited to $50,000 by statute.
- Qualified applicants: State and local governments
- Qualified technology: Biomass
- Administered by: National Institute of Food and Agriculture
- Authorization: National Agricultural Research, Extension, and Teaching Policy Act of 1977 (P.L. 95-113); Food, Conservation, and Energy Act of 2008 (P.L. 110-246)
- Annual funding: $875,000 for FY2010; an estimated $875,000 for FY2011; and an estimated $875,000 for FY2012
- Scheduled termination: Authorized through FY2012

- Description: This program provides grant funding for approved technology development, applied research, and training to develop an agriculture-based renewable energy workforce. The initiative shall support the following fields: (A) bioenergy; (B) pulp and paper manufacturing; and (C) agriculture-based renewable energy resources.
- Qualified applicants: Public or private nonprofit community colleges; advanced technology centers
- Qualified technologies: Biomass; bioenergy
- For more information: See CFDA program number 10.314 at https://www.cfda.gov

7. Repowering Assistance Program

- Administered by: Rural Development
- Authorization: Title IX, Section 9003 of the Farm Security and Rural Investment Act of 2002 (FSIRA, P.L. 107-171 is amended by Title IX, section 9001 of the Food, Conservation and Energy Act of 2008 (P.L. 110-246), creating new section 9004 under FSIRA
- Annual funding: Mandatory authorization: $35 million for FY2009, to remain available until expended.
- Discretionary authorization: $15 million authorized annually for FY2009-FY2012 ; $0 discretionary funding appropriated FY2009-FY2011
- Scheduled termination: Authorized through FY2012
- Description: The purpose of this program is to provide financial incentives to biorefineries in existence on June 18, 2008 to replace the use of fossil fuels used to produce heat or power at their facilities by installing new systems that use renewable biomass, or to produce new energy from renewable biomass.
- Qualified applicants: Eligible biorefinery. The biorefinery must have been in existence on or before June 18, 2008
- Qualified technologies: Renewable biomass
- For more Information: See program number 10.866 on the CFDA website https://www.cfda.gov and the USDA program website at http://www.rurdev.usda.gov/
- BCP_RepoweringAssistance.html;

8. Rural Energy For America Program (REAP) Grants and Loans

- Administered by: Rural Development
- Authority: Title IX, section 9006 of the Farm Security and Rural Investment Act of 2002 (FSIRA, P.L. 107-171) is amended by Title IX, section 9001 of the Food Conservation, and Energy Act of 2008 (P.L. 110-246), creating new section 9007 under FSIRA. The new section 9007 converted the federal Renewable Energy Systems and Energy Efficiency Improvements Program into the Rural Energy for America Program (REAP)
- Annual funding:
- Mandatory authorization: $55 million for FY2009; $60 million for FY2010; $70 million for FY2011; and $70 million for FY2012
- Discretionary authorization: $25 million authorized annually for FY2009-FY2012. Discretionary funding appropriated: $5 million for FY2009; $40 million for FY2010; and $40 million for FY2011, based on continuing resolution through 3/4/2011 (P.L. 111-322); FY2012 budget request was $37 million
- Scheduled termination: Authorized through FY2012
- Description: REAP promotes energy efficiency and renewable energy for agricultural producers and rural small businesses through the use of (1) grants and loan guarantees for energy efficiency improvements and renewable energy systems, and (2) grants for energy audits and renewable energy development assistance.
- Qualified applicants: Commercial; schools; state, local, and tribal governments; rural electric cooperative; agricultural; public power entities
- Qualified technologies: Solar water heat; solar space heat; solar thermal electric; photovoltaics; wind; biomass; hydroelectric; renewable transportation fuels; geothermal electric; geothermal heat pumps; CHP/cogeneration; hydrogen; direct-use geothermal; anaerobic digestion; small hydroelectric; tidal energy; wave energy; ocean thermal; renewable fuels; fuel cells using renewable fuels; microturbines. Specific energy efficiency technologies not identified
- For more information: See the program website at http://www.rurdev.usda.gov/rbs/farmbill/
-

9. Sustainable Agriculture Research and Education Program (SARE)

- Administered by: National Institute of Food and Agriculture; Agricultural Research Service; and other appropriate agencies
- Authorization: Food, Agriculture, Conservation and Trade Act of 1990 (P.L. 101-624); Food, Agriculture, Conservation and Trade Act Amendments of 1991 (P.L. 102-237); Federal Agriculture Improvement and Reform Act of 1996 (P.L. 104-127); Food, Conservation, and Energy Act of 2008 (P.L. 110-246)
- Annual funding: $12.5 million for FY2006; $12.4 million for FY2007; $9.1 million for FY2008; $14.5 million for FY2009; $14.5 million for FY2010, $15 million requested for FY2011
- Scheduled termination: None
- Description: The purpose of the Sustainable Agriculture Research and Education Program (SARE) is, in part, to encourage research designed to increase our knowledge concerning agricultural production systems that conserve soil, water, energy, natural resources, and fish and wildlife habitat. SARE provides grants through the agricultural bioenergy feedstock and energy efficiency research and extension initiative for projects with the purpose of enhancing the production of biomass energy crops and the energy efficiency of agricultural operations.
- Qualified applicants: Federal and state governments; colleges and universities; state agricultural experiment stations; state cooperative extension services; nonprofit organizations; individuals with demonstrable expertise
- Qualified technologies: Biomass; biofuels; other technologies not identified.
- For more information: See CFDA program number 10.215 at https://www.cfda.gov

IV. U.S. Department of Housing and Urban Development

1. Assisted Housing Stability and Energy and Green Retrofit Investments Program (Recovery Act Funded)

- Administered by: Department of Housing and Urban Development (HUD)
- Authority: American Recovery and Reinvestment Act of 2009 (ARRA; P.L. 111-5), Title XII
- Annual funding: (Project Grants) $0 for FY2009; $235 million for FY2010 ; $0 for FY2011—All obligations were to be made by September 30, 2010Scheduled termination: Anticipated to be 100% obligated and spent within two years of passage (by February 17, 2011). Receiving property owners are required to spend the funds on the specific improvements within two years of receipt
- Description: This program will provide funding for energy and green retrofit investments to certain eligible assisted, affordable multifamily properties. Funding includes incentives for participating property owners, a set-aside for administrative functions, and a set-aside for due diligence and underwriting support. Assistance will be for specific retrofit purposes.
- Qualified applicant: Residential
- Qualifying technologies: Specific technologies not identified
- For more information: See CFDA program number 14.318 at https://www.cfda.gov

2. Energy Efficient Mortgages (EEMs)

- Administered by: Federal Housing Administration (FHA) and Department of Veterans Affairs (VA). Conventional mortgages: Private lenders that sell mortgage loans to Fannie Mae or Freddie Mac may offer Energy Efficient Mortgages (EEMs)
- Authority: EEMs were initially introduced by lenders in the 1980s. In 1992, three pieces of legislation passed by Congress worked towards standardizing and expanding the use of EEMs. In 1992, Congress established an FHA Energy Efficient Mortgage Pilot Program (P.L.

102-550). The program was later expanded beyond five states to become a national program. The Housing Economic Recovery Act of 2008 (HERA; P.L. 110-289) increased the maximum amount that can be added to an FHA mortgage for energy efficient improvements. The 111th Congress also passed some incentives to encourage green home improvements in the American Recovery and Reinvestment Act of 2009 (ARRA; P.L. 111-5).

- Scheduled termination: None
- Description: Homeowners can take advantage of EEMs to finance a variety of energy efficiency measures, including renewable energy technologies, in a new or existing home. The U.S. federal government directly provides these loans through the FHA and VA lending programs. Fannie Mae and Freddie Mac will also purchase EEMs from primary lenders. Primary lenders may issue EEMs that do not conform to underwriting standards.
- Qualified applicants: The loan is available to anyone who meets the income requirements for FHA's Section 203 (b), provided the applicant can meet the monthly mortgage payments. New and existing owner-occupied homes of up to two units qualify for this loan. Cooperative units are not eligible. VA: available to qualified military personnel, reservists and veterans; Conventional: Applicants qualifying for a conventional mortgage are also eligible for an energy efficient mortgage
- Qualifying technologies: Passive solar space heat; solar water heat; solar space heat; photovoltaics; daylighting; and other technologies not specifically identified

For more information: See the HUD, Energy Star and DSIRE websites at http://www.hud.gov/ offices/hsg/sfh/eem/eemhome.cfm
http://www.energystar.gov/index.cfm?c=mortgages.energy_efficient_mortgages
http://www.dsireusa.org/incentives/incentive.cfm?Incentive_Code=US36F&re=1&ee=1V. Department of Veterans Affairs

1. Energy Efficient Mortgages (EEMs)

- Administered by: FHA and VA. Conventional mortgages: Private lenders that sell mortgage loans to Fannie Mae or Freddie Mac may offer EEMs
- Authority: EEMs were initially introduced by lenders in the 1980s. In 1992, three pieces of legislation passed by Congress worked towards standardizing and expanding the use of EEMs. In 1992, Congress established an FHA Energy Efficient Mortgage Pilot Program (P.L. 102-550). The program was later expanded beyond five states to become a national program. The Housing Economic Recovery Act of 2008 (HERA; P.L. 110-289) increased the maximum amount that can be added to an FHA mortgage for energy efficient improvements. The 111th Congress also passed some incentives to encourage green home improvements in the American Recovery and Reinvestment Act of 2009 (ARRA; P.L. 111-5)
- Scheduled termination: None
- Description: Homeowners can take advantage of EEMs to finance a variety of energy efficiency measures, including renewable energy technologies, in a new or existing home. The U.S. federal government directly provides these loans through the FHA and VA lending programs. Fannie Mae and Freddie Mac will also purchase EEMs from primary lenders. Primary lenders may issue EEMs that do not conform to underwriting standards.
- Qualified applicants: The loan is available to anyone who meets the income requirements for FHA's Section 203 (b), provided the applicant can meet the monthly mortgage payments. New and existing owner-occupied homes of up to two units qualify for this loan. Cooperative units are not eligible. VA: available to qualified military personnel, reservists and veterans; Conventional: applicants qualifying for a conventional mortgage are also eligible for an energy efficient mortgage
- Qualifying technologies: Passive solar space heat; solar water heat; solar space heat; photovoltaics; daylighting; and other technologies not specifically identified
- More information: See the HUD, Energy Star and DSIRE websites at http://www.hud.gov/offices/hsg/sfh/eem/eemhome.cfm; http://www.energystar.gov/index.cfm?c=mortgages.energy

http://www.dsireusa.org/incentives/incentive.cfm?Incentive_Code=US36F&re= 1&ee=1

VI. Small Business Administration

1. 7(a) Loan Guarantees

- Administered by: Small Business Administration (SBA)
- Authority: Small Business Act of 1953 (Public Law 83-163)
- Scheduled termination: None
- Annual Funding: $80 million ($10.2 million guaranty) for FY2010; $127.5 million ($83.2, million 7(a) guaranty, $41.2 7(a) guaranty revolvers and $3.1 million 7(a) dealer floor plan) for FY2011; and a $129.8 budget request ($87.5 million 7(a) guaranty, $39.2 million 7(a) guaranty revolvers, $3.1 million dealer floor plan) for FY2012.
- Description: To provide guaranteed loans from lenders to small businesses which are unable to obtain financing in the private credit marketplace, but can demonstrate an ability to repay loans if granted, in a timely manner. Guaranteed loans are made available to for-profit small businesses. The SBA's 7(a) lending authority includes (1) regular 7(a) (2) the Low Documentation Loan Program (Low Doc); (3) SBAExpress Program; (4) the Cap Line Program (5) PLP and (6) International Trade.
- Qualified applicant: Small businesses (meeting the size and eligibility standards)
- Qualified technologies: Not specifically listed
- For more information: See the SBA Website at http://www. sba. gov/financialassistance/borrowers and CFDA program number 59.012 at https://www.cfda.gov

VII. Department of the Interior

1. Energy and Mineral Development Program: Minerals and Mining on Indian Lands

- Administered by: Bureau of Indian Affairs. Energy and Mineral Development Program
- Authority: Indian Self-Determination and Education Assistance Act (P.L. 93- 638), 25 USC 450; Snyder Act of 1921 (P.L. 67-85), 25 USC 13; Indian Minerals Development Act (P.L. 97-382), 25 USC 2101 et seq.; Umatilla Basin Project Act (P.L. 101-557), 16 USC 1271 et seq.
- Annual funding: $18.622 million for FY2010; $19.998 million requested for FY2011
- Scheduled termination: None
- Description: Funding may be used to facilitate the inventory, assessment, promotion and marketing of both renewable and nonrenewable energy and mineral resources on Indian lands. Funds are awarded competitively to support assessment and inventory programs or to develop baseline data, but cannot be used for development purposes.
- Qualified applicants: Federally recognized Indian tribes; individual American Indian mineral owners
- Qualified technologies: Renewable energy technologies
- For more information: See CFDA program 15.038 at http:// www. cfda.gov

2. Tribal Energy Development Capacity Grant Program

- Administered by: Bureau of Indian Affairs
- Authorization: Energy Policy Act of 1992 (EPACT; P.L. 102-486); Tribal Energy Resource Development and Self-Determination Act of 2005 (Title V of Energy Policy Act of 2005; P.L. 109-58)
- Annual funding: $375,000 for FY2007, $1 million for FY2008; no estimate available for FY2009 or FY2010
- Scheduled termination: None
- Description: This program provides grants to Indian tribes to (1) develop and sustain the managerial and technical capacity needed to develop their energy resources; and (2) properly account for resulting energy production and revenues.
- Qualified applicant: Tribal governments
- Qualified technologies: Renewable energy technologies

- For information: See CFDA program number 15.148 at https://www.cfda.gov; or contact IEED, the Division of Indian Energy at (202) 219-0740

VIII. Department of Labor

1. Program of Competitive Grants for Worker Training and Placement in High Growth and Emerging Industry Sectors

- Administered by: Employment Training Administration
- Authority: American Recovery and Reinvestment Act of 2009 (ARRA; P.L. 111-5), Title VIII
- Annual funding: (Project Grants) $0 for FY2008; an estimated $750 million for FY2009; $0 for FY2010 and FY2011 Scheduled termination: None
- Description: This program provides competitive grants for worker training and placement in high growth and emerging industry sectors.
- Qualified applicants: State, local, and tribal governments; colleges and universities; private nonprofit institutions/organizations
- For more information: See the U.S. Department of Labor's (DOL's) *Training and Employment Notice* for this program at http://wdr.doleta.gov/directivesattach/ten/TEN44-08.pdf; and CFDA program number 17.275 at https://www.cfda.gov

IX. Department of Transportation

1. Hydrogen Storage Research and Development Program

- Administered by: Research Innovative Technology Administration (RITA)
- Authorization: Safe, Accountable, Flexible, Efficient, Transportation Equity Act: A Legacy for Users (SAFETEA-LU, P.L. 109-59)
- Annual funding: $213,878 for FY2009; $0 for FY2010
- Scheduled termination: June 30, 2011

- Description: This program provides grants for research and development of hydrogen storage technologies.
- Qualified applicant: Public nonprofit organizations; private nonprofit organizations
- Qualified technology: Hydrogen
- For more information: See CFDA program number 20.764 at https://www.cfda.gov

Appendix A. Summary of Federal Renewable Energy and Energy Efficiency Incentives/Index of Programs

Table A-1. Federal Incentives by Agency

Administering Agency	Program	Description	U.S. Code Citation	FY2010 Appropriations[a]	Expiration Date
Department of Energy	Advanced Research Projects Energy Financial Assistance Program (ARPA-E)	Grants to finance sophisticated energy technology R&D projects to accelerate transformation technology advances.	42 USC §16538	$0	Program evaluation after FY2012
	Biomass and Biorefinery Systems R&D Program	Grants to develop cost-effective technologies and systems to transform domestic biomass resources into biofuels, bioproducts, and biopower.	42 USC §16232	$220 million	None
	Building Technologies Program	Provides financial and technical assistance to improve efficiency of buildings and the equipment, components and systems within them	42 USC §17061-17124	$222 million	None
	Conservation Research and Development Grant Program	Grants to finance long-terms R&D efforts in buildings technologies, Industrial technologies, vehicle technologies, and hydrogen/fuel cell technologies.	42 USC §5901 et seq.	Estimated $1.7 billion	None

Administering Agency	Program	Description	U.S. Code Citation	FY2010 Appropriations[a]	Expiration Date
	Electricity Delivery and Energy Reliability, Research, Development and Analysis Grant Program	Grants to develop cost-effective technology to enhance the reliability, efficiency, and resiliency of the electric grid	42 USC §17381 et seq.	$125 million	None
	Energy Efficiency and Conservation Block Grants Program	Grants to finance energy efficiency and conservation programs/projects in local communities and renewable energy installations on government buildings	42 USC §17151-17158	$0	None
	Energy Efficiency and Renewable Energy Information Dissemination, Outreach, Training, and Technical Analysis/Assistance Program	Provides financial assistance to stimulate increased usage of energy efficiency/ renewable energy technologies and accelerate the adoption of these technologies	See Notes field[b]	Estimated $41.2 million	None

Table A-1. (Continued).

Administering Agency	Program	Description	U.S. Code Citation	FY2010 Appropriations[a]	Expiration Date
	Energy Efficiency and Renewable Energy Technology Deployment, Demonstration, and Commercialization Grant Program	Provides financial assistance for deployment, demonstration, and commercialization of energy efficiency and renewable energy technologies	42 USC §16191 et seq. and 42 USC §16231 et seq.	Estimated $7.2 million	None
	Energy Efficient Appliance Rebate Program	Provides financial and technical assistance to states to establish residential Energy Star rated appliance rebate programs	42 USC §15821	$0	End of FY2010
	Federal Energy Management Program	Provides assistance to federal agencies in developing and implementing energy efficiency and renewable energy technologies to meet energy management goals	42 USC §17131 et seq.	$32 million	None
	Financial Assistance Program (Office of Science)	Grants support research in the basic sciences and advanced technology concepts and assessments in fields related to energy	42 USC §13503	Estimated $1.3 billion	None

Administering Agency	Program	Description	U.S. Code Citation	FY2010 Appropriations[a]	Expiration Date
	Geothermal Technologies Program	Partners DOE with industry, academia, and research facilities to develop geothermal energy technologies	42 USC §16231 et seq. and 42 USC §17191 et seq.	$44 million	None
	Hydrogen & Fuel Cell Technologies Program	Partners DOE with industry, academia, and national laboratories to develop hydrogen and fuel cell technologies for the marketplace	42 USC §16151 et seq.	$174 million	None
	Industrial Technologies Program	Develops and supports the commercialization of new energy efficient technologies to improve industrial efficiency while increasing productivity	42 USC §17111 et seq.	$96 million	None
	Inventions and Innovations Program	Provides financial and technical assistance to develop innovative cost-effective ideas and inventions with future commercial value. Focus on energy efficiency and renewable energy technologies.	42 USC § 5913	Estimated $102,000	None

Table A-1. (Continued).

Administering Agency	Program	Description	U.S. Code Citation	FY2010 Appropriations[a]	Expiration Date
	Loan Guarantee Program	Loan guarantees to encourage commercial use of new or significantly improved technologies that avoid, reduce or sequester air pollutants or greenhouse gas emissions	42 USC §16511 et seq.	$0 for the Innovative Technology Loan Guarantee Program (Section 1703)	None
				$0 for the Temporary Loan Guarantee Program (Section 1705)	
	Regional Biomass Energy Programs	Provides financial assistance to increase America's use of fuels, chemicals, materials, and power made from domestic biomass	See Notes field[b]	Estimated $4.8 million	None
	Renewable Energy Production Incentive	Provides incentive payments for electricity generated and sold by new qualifying renewable energy facilities	42 USC §13317	$0	End of FY2026

Administering Agency	Program	Description	U.S. Code Citation	FY2010 Appropriations[a]	Expiration Date
	Renewable Energy Research and Development Program	Provides financial assistance to conduct R&D efforts in renewable energy technologies	42 USC §16231 et. seq.	Estimated $1.9 billion for FY2010 from ARRA funds	None
	Small Business Innovation Research/Small Business Technology Transfer Programs	Grants for small businesses to develop and commercialize energy technologies, including energy efficiency and renewable energy technologies	15 USC §638	$94 million	None
	Solar Energy Technologies Program	Program partners with industry, universities, and national laboratories to finance R&D and bring reliable and affordable solar energy technologies to the marketplace	42 USC §16231 et seq. and 42 USC §17171 et seq.	$247 million	None
	State Energy Program	Provides grants to states to design and implement their own renewable energy and energy efficiency programs	42 USC §6321 et seq.	$50 million	None
	Tribal Energy Program	Provides financial and technical assistance, education, and training to tribes to evaluate and develop renewable energy sources and energy efficiency measures	25 USC §3501 et seq.	$10 million	None

Table A-1. (Continued).

Administering Agency	Program	Description	U.S. Code Citation	FY2010 Appropriations[a]	Expiration Date
	Vehicle Technologies Program	Program partners with industry leaders to develop and deploy advanced transportation technologies to improve vehicle fuel efficiency and domestically produce clean and affordable alternative fuels	42 USC §17011 et seq.	$311.4 million	None
	Weatherization Assistance Program	Provides financial and technical assistance to states to increase the energy efficiency of low-income households	42 USC §6861 et seq.	$270 million	None
	Wind and Hydropower Technologies Program	Program partners with industry, states, federal entities, and other stakeholders on R&D projects to improve the performance, lower costs, and accelerate the deployment of wind and water power technologies	42 USC §16231 et. seq and 42 USC §17211 et seq.	$130 million	None

Administering Agency	Program	Description	U.S. Code Citation	FY2010 Appropriations[a]	Expiration Date
Internal Revenue Service	Business Energy Investment Tax Credit	Provides a tax credit for 30% of total expenditures on eligible systems placed in service, except geothermal systems, microturbines, and combined heat and power systems (10%)	26 USC §48	N/A	12/31/2016 for most eligible systems (except geothermal and solar thermal)
	Clean Renewable Energy Bonds (CREBs)	Bonds finance renewable energy projects	26 USC §54 (Old CREBs); 26 USC §54A (New CREBs) 26 USC §54C (New CREBs) IRS Notice 2009-33	N/A	12/31/2009 (Old CREBs); 11/01/2010 (New CREBS)

Table A-1. (Continued).

Administering Agency	Program	Description	U.S. Code Citation	FY2010 Appropriations[a]	Expiration Date
	Energy-Efficient Appliance Tax Credit for Manufacturers	Provides a tax credit to manufacturers for appliances that meet Energy Star 2007 requirements	26 USC §45M (amended)	N/A	12/31/2011
	Energy Efficient Commercial Buildings Tax Deduction	Tax deduction for certain qualifying systems and buildings	26 USC §179D (amended)	N/A	12/31/2013
	Energy-Efficient New Homes Tax Credit for Home Builders	Provides tax credits of up to $2,000 for builders of new, energy-efficient homes	26 USC §45L (amended)	N/A	12/31/2011
	Modified Accelerated Cost-Recovery System (MARCS) + Bonus Depreciation (2008-2012)	Allows businesses to recover investments in certain renewable energy property through depreciation deductions	26 USC §168 and 26 USC §48	N/A	5-year schedule for most solar, wind and geothermal. 12/31/2011 (100% bonus depreciation) and 12/31/2012 (50% bonus depreciation)

Administering Agency	Program	Description	U.S. Code Citation	FY2010 Appropriations[a]	Expiration Date
	Qualified Energy Conservation Bonds (QECBs)	Bond authority is allocated to state, local, and tribal governments to finance a broad range of energy efficiency and renewable energy projects	26 USC § 54A; 26 USC §54D; 26 USC § 6431	N/A	N/A
	Qualifying Advanced Energy Manufacturing Investment Tax Credit	Provides tax credits to encourage a U.S. based renewable energy manufacturing sector	26 USC §48C	N/A	Applications were due to DOE by 9/16/2009, with final applications due to DOE 10/16/2009.
	Renewable Energy Grants (1603 Program)	Renewable energy grant program to reimburse eligible taxpayers for a portion of the expense of placing in service specified energy property	ARRA § 1603(a)	N/A	Construction must begin by 12/31/2011 and Applications must be submitted before 10/1/2012.

Table A-1. (Continued).

Administering Agency	Program	Description	U.S. Code Citation	FY2010 Appropriations[a]	Expiration Date
	Renewable Energy Production Tax Credit (PTC)	Provides a per-kilowatt-hour tax credit for electricity generated by qualified renewable energy technologies and sold during the tax year	26 USC §45 (amended)	N/A	Generally, 10 years after the date the facility placed in service (with exceptions for some technology types)
	Residential Energy Conservation Subsidy Exclusion (Corporate)	Corporate tax exemption for energy-conservation subsidies are provided by public utilities, either directly or indirectly	26 USC §136 (amended)	N/A	None
	Residential Energy Conservation Subsidy Exclusion (Personal)	Personal tax exemption for energy-conservation subsidies provided by public utilities, either directly or indirectly	26 USC §136 (amended)	N/A	None
	Residential Energy Efficiency Tax Credit	Provides tax credit to residents/individuals for the installation of qualified energy efficient equipment to existing homes (primary residence)	26 USC §25C	N/A	12/31/2011

Administering Agency	Program	Description	U.S. Code Citation	FY2010 Appropriations[a]	Expiration Date
	Residential Renewable Energy Tax Credit	Provides a tax credit to residents/ individuals for the installation of qualified renewable energy systems to existing homes. Home must serve as owner's primary residence.	26 USC §25D (amended)	N/A	12/31/2016
	Alternative Motor Vehicle Credit	Provides tax credit for hybrid and lean-burn vehicles.	26 USC §30B	N/A	Varies by technology type: See Table A-2 below
Department of Agriculture	Assistance to High Energy Cost Rural Communities Program	Provides financial assistance to rural communities with high energy costs	7 USC. §918a	$18 million	None
	Bioenergy Program for Advanced Biofuels	Supports and ensures an expanding production of advanced biofuels by providing payments to advanced biofuels producers	7 USC §8105	$55 million	Authorized through FY2012
	Biomass Crop Assistance Program (BCAP)	Provides assistance to support the production of eligible biomass crops on land within approved project areas	7 USC §8111	"Such sums as necessary" with cap of $552 million	Authorized through FY2012

Table A-1. (Continued).

Administering Agency	Program	Description	U.S. Code Citation	FY2010 Appropriations[a]	Expiration Date
	Biorefinery Assistance Program	Assists in the development of new technologies for development of biofuels	7 USC §8103	$245 million	Authorized through FY2012
	Community Wood Energy Program	Provides grants to states and local governments to develop community wood energy plans or acquire or upgrade community wood energy systems	7 USC §8113	$0	Authorized through FY2012
	New Era Rural Technology Competitive Grants Program	Provides grant funding for approved technology development, applied research, and training to develop bioenergy and agriculture-based renewable energy resources	7 USC §3319e	$875K	Authorized through FY2012
	Repowering Assistance Program	Provides financial incentives to biorefineries in existence on June 18, 2008 to replace the use of fossil fuels used to produce heat or power by installing new systems that use renewable biomass or to produce new energy from renewable biomass	7 USC §8104	$0	Authorized through FY2012

Administering Agency	Program	Description	U.S. Code Citation	FY2010 Appropriations[a]	Expiration Date
	Rural Energy for America Program	Provides grants and loan guarantees to promote energy efficiency and renewable energy to agricultural producers and rural small businesses	7 USC §8107	$100 million ($60 million mandatory and $40 million discretionary)	Authorized through 2012
	Sustainable Agriculture Research and Education	Provides grants for research projects with the purpose of enhancing biomass energy crop production and increasing the energy efficiency of agricultural operations	7 USC §5801 et seq.	$14.5 million	None
Department of Housing and Urban Development	Assisted Housing Stability and Energy Green Retrofit Investments Program	Provides grants or loans for energy retrofit and green investments in assisted, affordable multifamily housing	See Notes field[b]	$235 million	All obligations were to be made by 9/30/2010.
	Energy Efficient Mortgages	Provides backing of loans for energy efficient mortgages to finance the installation of energy efficiency or renewable energy technologies in new or existing homes	12 USC §1701z-16	N/A	None

Table A-1. (Continued).

Administering Agency	Program	Description	U.S. Code Citation	FY2010 Appropriations[a]	Expiration Date
Department of the Interior	Energy and Mineral Development Program: Minerals and Mining on Indian Lands	Facilitate the inventory, assessment, promotion and marketing of both renewable and nonrenewable energy and mineral resources on Indian lands	25 USC §450; 25 USC §13; 25 USC §2101 et seq; 16 USC. §1271 et seq.	$18.622 million	None
	Tribal Energy Development Capacity Grant	Grants to Indian tribes to develop and sustain the managerial and technical capacity needed to develop their energy resources and properly account for resulting energy production and revenues	25 USC §3502	No estimate available	None
Department of Labor	Program of Competitive Grants for Worker Training and Placement in High Growth and Emerging Industry Sectors	Intended to preserve and create jobs; promote economic recovery; assist those most impacted by the recession; provide investments and invest in infrastructure	See Notes fieldb	$0	None

Administering Agency	Program	Description	U.S. Code Citation	FY2010 Appropriations[a]	Expiration Date
	Small Business Administration	7(a) Loan Guarantees	Provides guaranteed loans from lenders to small businesses	15 U.S.C. 636(a)(25(B)	$80 million
Department of Transportation	Hydrogen Storage Research and Development Program	This program provides grants for research and development of hydrogen storage technologies	42 USC §128	$0	6/30/2011
Department of Veterans Affairs	Energy Efficient Mortgages	Provides backing of loans for energy efficient mortgages to finance the installation of energy efficiency or renewable energy technologies in new or existing homes	12 USC §1701z-16	N/A	None

Source: CRS.

a. FY2010 Appropriations data compiled from executive agency budget justifications, congressional committee reports, and program descriptions from the online edition of the *Catalog of Federal Domestic Assistance*.

b. Some programs are not specifically identified or codified in the United States Code.

Table A-2. Alternative Motor Vehicle Credit (26 USC §30B)

Type of Credit	Expiration Date
Fuel Cell Motor Vehicle Credit	12/31/2014
Advanced Lean Burn Technology Motor Vehicle Credit	12/31/2010
Qualified Plug-In Electric Drive Motor Vehicle Credit	12/31/2014
Qualified Plug-In Electric Motor Vehicle Conversion Credit	12/31/2011
Qualified Alternative Fuel Motor Vehicle Credit	12/31/2010
Qualified Hybrid Motor Vehicle Credit	12/31/2010

Source: U.S. Code and IRS.

Appendix B. Index of Programs by Applicant Eligibility and Technology Type

Table B-1. Index of Programs by Applicant Eligibility

Applicant Eligibility	Program Numbers
Advanced Technology Centers	III-6
Agricultural/Extension/Biofuel Producers	II-6, III-2, III-3, III-4, III-7, III-8, III-9
Alaska Native Corporations	I-15
Builder/Developer	II-4, II-5
Commercial/Industrial/For-Profit	I-1, I-2, I-3, I-4, I-5, I-6, I-11, I-12, I-13, I-15, I-16, I-18, I-21, I-22, I-24, I-25, II-3, II-4, II-6, II-7, II-8, II-9, II-14, III-1, III-2, III-3, III-4, III-7, III-8
Cooperative/Collaborative/Consortia	I-21, II-12, III-1, III-4, III-8
Federal Government	I-4, I-6, I-12, I-23, II-4, III-9
Higher Education (Colleges and Universities)	I-1, I-2, I-3, I-4, I-5, I-6, I-7, I-12, I-13, I-15, I-18, I-21, I-22, I-24, I-25, III-4, III-6, III-9, VIII-1
Local Government	I-2, I-6, I-7, I-12, I-13, I-14, I-15, I-17, I-18, I-22, I-24, I-25, II-12, II-13, III-1, III-4, III-5, III-8, VIII-1
National Laboratories	I-4, I-5, I-6, I-7, I-12, III-4

Applicant Eligibility	Program Numbers
Nonprofit	I-2, I-13, I-15, I-17, I-18, I-21, I-22, I-24, I-25, III-1, III-9, VIII-1, IX-1
Other/Cross-Cutting	I-21, II-14, II-15
Research Organization	I-21, I-22
Residential/Individual	I-10, I-15, II-1, II-2, II-10, II-11, II-15, III-1, III-4, III-9, IV-1, IV-2, V-1
Schools	III-8
Small Businesses	I-6, I-10, I-24, I-26, VI-1
State Government	I-2, I-6, I-7, I-8, I-9, I-12, I-13, I-14, I-15, I-16, I-17,I-18, I-19, I-22, I-24 , I-25, II-4, II-12, II-13, III-1, III-4, III-5, III-8, III-9, VIII-1
Tribal Government	I-6, I-9, I-13, I-14, I-15, I-17, I-18, I-19, I-20, I-22, I 24, I-25, II-12, II-13, III-1, III-4, III-8, VII-1, VII-2, VIII-1
U.S. Territories	I-8, I-9, I-14, I-19,
Utilities	I-17, II-3, II-12, III-4, III-8
Veterans	IV-2, V-1

Source: CRS.

a. Program numbers correspond to agency (roman numeral) and number assigned to each program within this document. See Table of Contents.

Table B-2. Index of Programs by Technology Type

Qualified Technologies	Program Numbers[a]
Advanced Batteries	I-13,
Air Conditioners	I-20, II-1, II-4
Alternative Vehicles/Vehicle Technologies	I-12, II-8, II-14, II-15
Anaerobic Digestion	I-17, II-6, II-9, II-12, II-13, II-14, III-8
Appliances (Energy Efficient) (All)	I 8
Batteries (Energy Storage)	II-8
Biodiesel / Biofuels	I-12, I-25, III-2, III-4, III-9
Boilers	I-20, II-1, II-4

Table B-2. (Continued).

Qualified Technologies	Program Numbers[a]
Biomass	I-1, I-2, I-14, I-16, I-17, I-18, I-20, I-23, II-1, II-3, II-6, II-9, II-
	12, II-13, II-14, III-2, III-3, III-4, III-5, III-6, III-7, III-8, III-9
Caulking/Weather Stripping	I-9, I-20, II-4
Chillers	I-20, II-4
Clothes Washers	I-20, II-7
Combined Systems/CHP/Energy Management Systems	I-7, I-14, I-20, II-3, II-6, II-14, III-8
Comprehensive/Whole Building	I-20, II-4, II-5
Dishwashers	II-7
Doors	I-20, II-1, II-4
Duct/Air Sealing	I-9, I-20, II-4
Equipment	I-7, I-14
Fuel Cells	I-4, I-7, I-13, I-14, I-16, I-18, I-25, II-1, II-2, II-3, II-6, II-8, II-14,
	II-15, III-8
Furnaces	I-20, II-1, II-4
Geothermal (All)	I-3, I-16, I-18, I-23, II-3, II-14, III-8
—Geothermal (Direct Use)	II-3, II-14, III-8
—Geothermal (Electric)	I-17, I-20, I-25, II-3, II-6, II-8, II-9, II-12, II-13, II-14, III-8
—Geothermal (Heat Pumps)	I-20, II-2, II-3, II-6, II-8, II-14, III-8
Heat Pumps	II-1, II-4
Hybrid Electric	I-12, II-15
Hydrogen	I-4, I-13, I-16, I-18, II-9, III-8, IX-1
Hydropower (All)	I-6, I-16, I-18, I-23, II-6, II-9, II-12, II-13
—Hydroelectric	I-6, I-20, I-25, II-6, II-9, II-12, II-13, III-8
—Hydrokinetic	I-6, II-6, II-9, II-12, II-13
—Ocean	I-6, I-17, I-23, I-25, II-6, II-9, II-12, II-13, III-8
—Tidal	I-6, I-17, I-25, II-6, II-9, II-12, II-13, III-8
Qualified Technologies	Program Numbers[a]

—Wave	I-6, I-17, I-25, II-6, II-9, II-12, II-13, III-8
Insulation	I-9, I-20, II-1, II-4,
Landfill Gas	I-17, II-6, II-9, II-12, II-13, II-14
Lighting/Lighting Sensors	I-7, I-14, I-20, I-25, II-3, II-4, II-6, II-8, II-14, IV-2, V-1
Manufacturing Facilities	I-25
Microturbines	II-3, II-6, II-8, II-14, III-8
Municipal Solid Waste	II-6, II-9, II-12, II-13, II-14
Other Technologies[b]	I-9, I-10, I-11, I-15, I-19, I-20, I-21, I-22, I-24, I-26, II-8, II-10, II-11, III-1, III-8, III-9, IV-1, IV-2, V-1, VI-1, VII-1, VII-2
Programmable Thermostats	I-20
Refrigerators/Freezers	I-20, II-7
Renewable Transportation Fuels	I-25, II-14, III-8
Roofs	I-20, II-1, II-4
Siding	I-20, II-4
Smart Grid	II-8
Solar (All)	I-5, I-7, I-14, I-16, I-18, I-23, II-2, II-3, II-6, II-14, III-8
—Photovoltaics	I-7, I-17, I-20, I-25, II-2, II-3, II-6, II-8, II-10, II-11, II-12, II-13, II-14, III-8, IV-2, V-1
—Solar Space Heat	I-20, II-2, II-3 , II-6, II-10, II-11, II-14, III-8, IV-2, V-1
—Solar Thermal Electric/Process	I-17, I-25, II-2, II-3, II-6, II-8, II-12, II-13, II-14, III-8
—Solar Water Heat	II-2, II-3, II-6, II-8, II-10, II-11, II-14, III-8, IV-2, V-1
Water Heaters	I-20, II-1, II-4
Wind	I-6, I-14, I-17, I-18, I-20, I-23, I-25, II-2, II-3, II-6, II-8, II-9, II-12, II-13, II-14, III-8
Windows	I-7, I-9, I-20, II-1, II-4

Source: CRS.

[a] Program numbers correspond to agency (roman numeral) and number assigned to each program within this document. See Table of Contents.

[b] Other technologies includes cross-cutting and advanced technologies, other unspecified technologies, all energy efficiency and/or renewable energy technologies, or not specifically identified.

End Notes

[1] For more information, see page 500 in volume three of the Department of Energy's *FY 2011 Congressional Budget Request* at http://www.cfo.doe.gov/budget/11budget/Content/Volume 3.pdf.

[2] For more information, see p. 30 of CRS Report R40669, *Energy and Water Development: FY2010 Appropriations*, coordinated by Carl E. Behrens.

In: Federal Incentives for Renewable Energy ISBN: 978-1-61324-946-8
Editors: L.Belisario and N. Meier

Chapter 2

RENEWABLE ENERGY AND ENERGY EFFICIENCY TAX INCENTIVE RESOURCES*

Lynn J. Cunningham and Beth A. Roberts

SUMMARY

The following list of authoritative resources is designed to assist in responding to a broad range of constituent questions and concerns about renewable energy and energy efficiency tax incentives. Links are provided for the following: the full text of public laws establishing and extending federal renewable energy and energy efficiency incentives; federal, state, and local incentives resources; incentive resources grouped by technology type (solar, wind, geothermal, and biomass); and federal grants information resources. The last section of this report includes tables displaying popular incentives, the corresponding *U.S. Code* citations, and current expiration dates of those incentives.

* This is an edited, reformatted and augmented version of a Congressional Research Service publication, CRS Report for Congress R40455, from www.crs.gov, dated March 23, 2009.

Full Text of Tax Incentive Legislation

Energy Policy Act of 2005 (EPACT) P.L. 109-58. The tax provisions are located in Title XIII.

Emergency Economic Stabilization Act of 2008 (EESA) P.L. 110-343. The tax provisions are located in Titles I, II, and III of Division B, the Energy Improvement and Extension Act of 2008.

American Recovery and Reinvestment Act of 2009 (ARRA) P.L. 111-5. Conference Report with full text of the act (H.R. 1, as passed, P.L. 111-5) and the Joint Explanatory Statement. The tax provisions are located in Division B, Title I.

Federal Incentives

This section lists and describes several resources that contain information about federal incentives available to support energy efficiency and renewable energy.

Tax Incentives Assistance Project (TIAP)
http://www.energytaxincentives.org/

This website is sponsored by a number of government agencies, nonprofit groups, and other organizations. It focuses solely on information about federal tax incentives. Information is organized into categories for consumers, businesses, and builders/manufacturers. The site includes updates about enacted federal legislation and provides links to Internal Revenue Service (IRS) tax forms.

Environmental Protection Agency (EPA) Energy Star
http://www.energystar.gov/index.cfm?c=products.pr_tax_credits

This website has a page on "Federal Tax Credits for Energy Efficiency." The information on that page is organized into categories for consumers (home improvements, cars, solar energy, fuel cells), home builders, appliance manufacturers, and commercial buildings. The site includes a frequently asked questions (FAQ) section providing answers about energy efficiency tax credits.

Department of Energy (DOE) Financial Opportunities

http://www1.eere.energy/

This website is focused mainly on information about matching funds, grants, and financing. Information is organized into categories for consumers, business/industry/universities, inventors (small business), federal energy managers, states, and Native American tribes. The site includes a section on energy efficiency and consumer home financing.

U.S. Department of Energy Alternative Fuels and Advanced Vehicles Data Center (AFDC)

http://www.eere.energy

This website presents information about incentives for alternative fuels (renewable fuels and others) and vehicles. A key link provides access to "State and Federal Incentives and Laws." Incentives covered include grants, tax credits, loans, rebates, regulatory exemptions, fuel discounts, and technical assistance. Information on state incentives is made available through a national map and through summary tables organized by type of incentive, regulation, technology/fuel, and user. The information about state incentives is updated after each state legislature's session ends. Information about federal incentives is updated after pertinent legislation is enacted into law. Another link provides access to "Laws and Incentives Enactment History."

U.S. Department of Energy (DOE) Clean Cities Financial Opportunities

http://www1.eere.energy

This website presents information about incentives for alternative fuels and advanced technologies. A link to "Government Sources" provides information about funding opportunities through federal grant-making agencies (Grants.gov), Metropolitan Planning Organization (MPO), the Congestion Mitigation and Air Quality (CMAQ) Program, and various EPA programs. A link to "Solicitations" provides information about business funding opportunities that cover a variety of changing topics that have included plug-in hybrid vehicles, hydrogen vehicles, and transportation planning. Clean Cities Coordinators are available to help with funding applications.

Department of Housing and Urban Development (HUD) Energy Efficient Mortgages Program

http://www.hud.gov/offices/hsg/sfh/eem/energy

HUD's website provides information on Energy Efficient Mortgages. These mortgages can help homeowners finance the cost of adding energy-efficiency features to new or existing housing as part of their Federal Housing Authority-insured home purchase or refinance.

Alliance to Save Energy (ASE) Home and Vehicle Tax Credits

http://www.ase.org/content/article/detail/2654

ASE's website organizes information into categories on energy efficiency incentives for home improvements, hybrid vehicles, and solar energy. The site includes details on eligible equipment, credit limits, and credit expiration dates.

State and Local Incentives

This section covers websites that list and describe state and local incentives available to support energy efficiency and renewable energy.

Database of State Incentives for Renewables and Efficiency (DSIRE) http://www.dsireusa.org/

DSIRE's website is sponsored by the Interstate Renewable Energy Council (IREC). IREC is a nonprofit organization focused on standards, guidelines, and other activities to support renewable energy. This site contains information about various types of energy efficiency and renewable energy financial incentives provided by state and local governments and utility companies. Summary data is accessed through a national map—and several additional special topic maps— that are linked to data on each state. Alternatively, the data can be searched by technology (solar, wind, geothermal), sector (residential, commercial/industrial, government, utility), and incentive type (tax credits, bonds, grants, loans), and eligible and implementing sectors. The site is updated weekly. Also, the homepage includes a map-link to a list of federal incentives.

United States Department of Energy (DOE) State Activities and Weatherization Assistance

http://www.eere.energy

DOE's website on the Weatherization Assistance Program has information about how to apply for weatherization funding assistance. The site also has a "state activities" link, which provides information about state-level energy assistance programs.

Incentives by Technology Type

The following are links to resources by type of renewable energy.

Biomass

Database of State Incentives for Renewables and Energy Efficiency (DSIRE). Incentive for Biomass

http://www.dsireusa.org/

Energy Efficiency and Renewable Energy. Alternative Fuels and Advanced Vehicles Data Center. State and Federal Incentives and Laws

http://www.eere.energy

Environmental Protection Agency. Funding Database Biomass/Biogas

http://www.epa.gov/chp/funding

Geothermal

Database of State Incentives for Renewables and Energy Efficiency (DSIRE). Incentives for Geothermal Heat Pumps and Geothermal Electric

http://www.dsireusa.org/

Geothermal Heat Consortium – GeoExchange.org. New Federal Tax Credits for Geothermal Heat Pump Systems

http://www.geoexchange.org/component/content/article/90-new-federal-tax-credits-.html

Solar

Alliance to Save Energy. Energy-Efficiency Home and Vehicle Tax Credits. Solar Energy and Fuel Cell
http://www.ase.org/content/article/detail/2654#fuelcells_solar

Database of State Incentives for Renewables and Energy Efficiency (DSIRE). Incentives for Solar Technology
http://www.dsireusa.org/

Energy Star. Federal Tax Incentives for Renewable Energy. Tax Incentives for Solar Energy Systems
http://www.energystar.gov/index.cfm?c=products.pr_tax_credits#s4

Solar Energy Industries Association. Solar Bills/Legislation http://www.seia.org/cs/solar_bills

Solar Energy Industries Association. Frequently Asked Questions on the Solar Investment Tax Credit http://www.seia.org/galleries/pdf/ITC Frequently Asked Questions 10 9 08 .pdf

Tax Incentives Assistance Project. Consumer Tax Incentives. Solar Energy Systems http://www.energytaxincentives.org/consumers

Tax Incentives Assistance Project. Businesses Tax Incentives. Solar Energy Systems http://www.energytaxincentives.org/business/renewables.php

Wind

American Wind Energy Association. Legislative Affairs

http://www.awea.org/legislative/

Database of State Incentives for Renewables and Energy Efficiency (DSIRE). Incentives for Wind

http://www.dsireusa.org

POPULAR INCENTIVES TABLES

Table 1. U.S. Code Citations and Expiration Dates for Popular Renewable Energy an Energy Efficiency Tax Incentives/Credits

Type of Credit	U.S. Code Citation	Expiration Date
Biofuels Production Tax Credit (PTC)	26 U.S.C. § 40A	12/31/2009
Business Solar and Fuel Cell Investment Tax Credit (ITC)	26 U.S.C. § 48(a)(3)	12/31/2016 [a]
Energy Efficiency Appliance Credit	26 U.S.C. § 45M	12/31/2008-12/31/2010
Energy Efficient Commercial Buildings Deduction	26 U.S.C. § 179D	12/31/2013
Energy Efficient New Homes Tax Credit for Homebuilders	26 U.S.C. § 45L	12/31/2009
Renewable Energy Production Tax Credit (PTC)	26 U.S.C. § 45	Wind (12/31/2012) Biomass, Geothermal, Hydroelectric, and Marine and Hydrokinetic (12/31/2013)
Residential Energy Efficient Property Credit (Solar	26 U.S.C. § 25D	12/31/2016

Source: U.S. Code

[a] Depending on efficiency level of appliance.

Table 2. Alternative Motor Vehicle Credit

Type of Credit	U.S. Code Citation	Expiration Date
Advanced Lean Burn Technology Motor Vehicle Credit	26 U.S.C. § 30B	12/31/2010
Fuel Cell Motor Vehicle Credit	26 U.S.C. § 30B	12/31/2014
New Qualified Plug-In Electric Drive Motor Vehicles	26 U.S.C. § 30B	12/31/2014
Qualified Plug-In Electric Motor Vehicle Conversion Credit	26 U.S.C. § 30B	12/31/2011
Qualified Alternative Fuel Motor Vehicle Credit	26 U.S.C. § 30B	12/31/2010
Qualified Hybrid Motor Vehicle Credit	26 U.S.C. § 30B	12/31/2009

Source: U.S. Code.

GRANTS INFORMATION

Catalog of Federal Domestic Assistance (CFDA)

http://www.cfda.gov/

The CFDA is the primary source of federal grants program information, although actual funding depends upon annual congressional budget appropriations. The CFDA is available on the Internet. Many federal grants do not provide funding directly to individuals, but rather to states, local governments, universities, and tribal entities. Check the "applicant eligibility" section in the CFDA program description to see who may apply. Individuals may be eligible to apply for funds after they have been distributed at the state and local level, through their state energy offices or other contact listed in the CFDA program description.

Grants.gov

http://www.grants

Federal grant funding opportunities are also posted on the website Grants.gov. Grants.gov enables grant seekers to electronically find and apply for competitive grants from all federal agencies.

In: Federal Incentives for Renewable Energy ISBN: 978-1-61324-946-8
Editors: L.Belisario and N. Meier

Chapter 3

ARRA SECTION 1603 GRANTS IN LIEU OF TAX CREDITS FOR RENEWABLE ENERGY: OVERVIEW, ANALYSIS, AND POLICY OPTIONS*

Phillip Brown and Molly F. Sherlock

SUMMARY

Congress created the Section 1603 grant program as part of the American Recovery and Reinvestment Act of 2009 (ARRA; P.L. 111-5). This program, administered by the U.S. Department of the Treasury, provides cash grant incentives for renewable energy projects. Initially, the Section 1603 grant program was scheduled to expire at the end of 2010. A one-year extension was enacted as part of the Tax Relief, Unemployment Reauthorization, and Job Creation Act of 2010 (P.L. 111-312) at an estimated cost of $3 billion. Absent Congressional action, the Section 1603 grant program will expire at the end of 2011.

As of December 6, 2010, grants totaling approximately $5.6 billion had been awarded to 1,495 entities, since Section 1603 became law in February 2009. Wind has received approximately 84% of the grant award value, while solar electric represents approximately 75% of entities that have received grant awards. "Other" technologies (qualifying energy

---

* This is an edited, reformatted and augmented version of a Congressional Research Service publication, CRS Report for Congress R41635, from www.crs.gov, dated February 8, 2011.

property not represented by wind or solar electricity) have also received grant awards, although the value and number of awards represented by this category is relatively small compared to wind and solar electricity.

Prior to the availability of Section 1603 grants, qualifying renewable energy projects were federally supported primarily through the production tax credit (PTC) or investment tax credit (ITC). It has been common industry practice for renewable energy developers to partner with tax-equity investors, where the tax-equity investors offer cash in exchange for project ownership, project cash flows, tax credits, and depreciation benefits. The Section 1603 grant program was motivated by difficult economic conditions and the perceived lack of tax-equity capacity to support renewable energy projects. Analysis of the tax equity marketplace reveals fluctuations in the dollar volume, number of participants, and required rates of return between 2007 and 2010.

Market response, since Section 1603 was established, has been mixed. The solar industry expects a record year of installations in 2010 (approximately 1 Gigawatt), while the wind industry forecasts a 50% decline compared to 2009 (approximately 5 Gigawatts installed in 2010 compared to 10 Gigwatts installed in 2009). It is important to note, however, that many factors influence annual renewable energy installations, the cash grant being just one.

If Congress considers additional extensions to or modifications of the Section 1603 grant program, economic and cost factors may also be taken into account. Grants, as opposed to tax credit, may be a more efficient mechanism for delivering public funds to the renewable energy sector. As is the case with most tax or subsidy programs, however, there are concerns that grants may be going to projects that would have moved forward without added federal incentives.

Finally, this report presents various policy options Congress may want to consider regarding the Section 1603 grant and related tax credits for renewable energy. The first option presented is to allow the grant program to expire. Even if the grant program were to expire, tax incentives would remain available. A second option is to extend the Section 1603 grant program. An extension of the grant program may be considered alongside and extension of the PTC for wind, which is set to expire at the end of 2012. A modification to the ITC and PTC, that could potentially enhance the benefits associated with the existing tax incentives, is presented as a third option.

INTRODUCTION

Broad efforts are underway to expand the use of renewable energy technologies to increase domestic energy production, improve U.S. energy security, create new industries and new jobs, and to reduce greenhouse gas emissions. According to the U.S. Energy Information Administration, renewable energy technologies provided 3.6% of U.S. electricity generation in 2009. Of the 144 million megawatt hours of electricity generated from renewable sources, wind provided 51.4%, wood and wood-derived fuels provided 24.8%, other forms of biomass provided 12.8%, geothermal accounted for 10.4%, and solar thermal/photovoltaic provided 0.6%.[1] Although U.S. renewable electricity markets have grown in size and scope in recent years, the cost of renewable electricity generation continues to be typically higher than conventional generation sources (e.g. coal, natural gas, existing nuclear). Federal tax policy has served to reduce this cost gap by providing various tax incentives for renewable energy.

Tax incentives, historically, have promoted investment in renewable electricity generation by rewarding either electricity production or investment in renewable generation capacity. In recent years, the size of tax benefits available for renewable energy investors has sometimes exceeded the investor's tax liability. Investors may carry unused tax credits forward to offset future tax liability, or, alternatively, partner with a third-party tax-equity investor capable of providing cash in exchange for tax benefits.

In February of 2009, Congress passed new legislation that provides renewable energy projects with an alternative option for realizing the value of federal tax benefits: a grant in lieu of tax credits (Section 1603 of the American Recovery and Reinvestment Act of 2009 (P.L. 111-5)). Commonly known as "Section 1603," this program affords the project the opportunity to receive a one-time cash grant instead of tax credits, thereby reducing the need for third-party tax-equity investors. One key motivation for creating the Section 1603 grant program was the perceived lack of tax equity capacity available for renewable electricity generation projects following the financial crisis in 2007. The Section 1603 grant program was initially enacted on a temporary basis, and was scheduled to expire at the end of 2010. Congress extended the program for one year, through 2011, in the Tax Relief, Unemployment Reauthorization, and Job Creation Act of 2010 (P.L. 111-312).

This report provides a comprehensive overview of the Section 1603 grant program. The first part of this report focuses on how different renewable energy market sectors have responded since the grant option was made available. Understanding how different renewable energy markets have responded to the grant option may provide some insight regarding the potential implications for the renewable energy market of letting the Section 1603 grant program expire as scheduled at the end of 2011.

The Section 1603 grant program was established in response to weak tax-equity markets. This report provides a detailed overview of tax-equity markets during the recent financial crisis, as well as information on the current status of tax-equity markets. Since weak tax-equity markets were a motivating factor in adopting the Section 1603 grant program in 2009, understanding trends in tax-equity markets may help Congress evaluate whether the policy motivations that led to the adoption of the Section 1603 grant program under ARRA remain relevant.

An economic perspective may also prove useful in evaluating a possible extension of the Section 1603 grant program. Oftentimes, subsidies that encourage the use of renewable energy resources are not the most economically efficient means of achieving a policy objective. Compared to direct tax incentives, however, the Section 1603 grant program may be an improvement in terms of economic efficiency.

The final sections of this report provide various policy options related to the Section 1603 grant program. As noted above, absent Congressional action, the Section 1603 grant program is scheduled to expire at the end of 2011. One option is to further extend the Section 1603 grant option. As another option, existing tax incentives could be modified to provide some of the added benefits associated with the Section 1603 grant program.

BACKGROUND: SECTION 1603 GRANTS IN LIEU OF TAX CREDITS

Section 1603 of the American Recovery and Reinvestment Act of 2009 (ARRA; P.L. 111-5) provides grants for investments in certain energy production property in lieu of tax credits available under §§45 and 48 of the Internal Revenue Code (IRC). Table 1 summarizes the available incentives by eligible technologies. Prior to ARRA, investments in renewable energy technologies were generally eligible for the production tax credit (PTC (IRC §45)) or investment tax credit (ITC (IRC §48)), depending on technology type. ARRA modified the incentives available for renewable energy investments in two ways: (1) PTC-eligible property was given the option to claim the ITC at the ITC rate, and (2) all eligible energy properties were allowed to elect to receive a one-time cash grant in lieu of tax credits.

The Section 1603 grant program provides payments equal to 10% or 30% of the eligible cost basis for specified energy projects. Under ARRA, projects placed in service during the 2009 or 2010 tax years, or those that met certain start-of-construction requirements before the end of 2010, were eligible for the grant option.[2] The Tax Relief, Unemployment Reauthorization, and Job Creation Act of 2010 (P.L. 111-312) extended the placed-in-service and construction start dates through the end of 2011. Qualifying projects placed in service, as well as those that meet start-of-construction requirements during 2011, and that are placed in service by the deadline noted in Table 1, may elect to receive a Section 1603 grant.

Renewable Energy Tax Incentives: Pre-2009

As was noted above, before 2009, the ITC and PTC were the predominant tax incentives for renewable energy.[3] The technology receiving the most federal subsidization under the ITC is solar (PV, solar thermal, etc.) electricity, where qualified investments are eligible for a 30% tax credit. Table 1 summarizes the ITC rates available for other technologies and well as the ITC placed-inservice deadlines.[4] The PTC provides a per kilowatt-hour tax credit for renewable energy produced at qualified facilities.[5] Generally, the PTC is available for 10 years following an eligible facility's being placed-in-service. Table 1 summarizes the current PTC rates for various eligible technologies as well as the placed-in-service deadlines.

The PTC has been the primary incentive for wind since its adoption in 1992.[6] Since 1992, the PTC has been allowed to lapse a number of times.[7] Historically, these lapses have been followed by a short-term extension of the incentive, allowing the program to again lapse in a few years. Annual installations in wind capacity are much lower in years where the PTC has been allowed to lapse, suggesting that the credit plays an important role in wind deployment.[8]

Table 1. PTC and ITC/1603 Incentives for Renewable Energy Property

Energy Property	Required In-Service Date	2010 PTC Rate ($/kWh)	ITC/1603: % of Eligible Cost Basis	Incentive options before ARRA	Incentive options after ARRA
Wind	1/1/2013	$0.022	30%	PTC	PTC, ITC, Grant
Closed-Loop Biomass	1/1/2014	$0.022	30%	PTC	PTC, ITC, Grant
Open-Loop Biomass	1/1/2014	$0.011	30%	PTC	PTC, ITC, Grant
Geothermal Energya	1/1/2014	$0.022	30%	PTC	PTC, ITC, Grant
Landfill Gas	1/1/2014	$0.011	30%	PTC	PTC, ITC, Grant
Trash	1/1/2014	$0.011	30%	PTC	PTC, ITC, Grant
Qualified Hydropower	1/1/2014	$0.011	30%	PTC	PTC, ITC, Grant
Marine & Hydrokinetic	1/1/2014	$0.011	30%	PTC	PTC, ITC, Grant
Small Windb	1/1/2017	n.a.	30%	ITCb	ITC, Grant
Solar	1/1/2017	n.a.	30%	ITC	ITC, Grant
Geothermal Heat Pumps	1/1/2017	n.a.	10%	ITC	ITC, Grant
Fuel Cellsc	1/1/2017	n.a.	30%	ITC	ITC, Grant
Microturbinesd	1/1/2017	n.a.	10%	ITC	ITC, Grant
Combined Heat & Power (CHP)	1/1/2017	None	10%	ITC	ITC, Grant

Source: IRC §§45 and 48; U.S. Department of the Treasury 1603 guidance; IRS Notice 2010-37.

[a] Geothermal energy property qualifies for a 10% ITC under IRC § 48. As PTC-eligible property, geothermal energy investments are eligible for a 30% grant under Section 1603 of ARRA.

[b] Small wind turbines are those with a up to 100 kW in capacity. Prior to ARRA, small wind was subject to a $4,000 credit limit.

[c] Maximum payment for fuel cell property may not exceed $1500 for each 0.5 kilowatt of capacity.

[d] Maximum payment for microturbine property may not exceed $200 for each kilowatt of capacity.

Investments in renewable energy also benefit from favorable depreciation schedules. Under the Modified Accelerated Cost Recovery System (MACRS), a number of renewable energy technologies are classified as five-year property, including solar and wind.[9] Generally, taxpayers are required to capitalize investments in capital assets, and deduct through depreciation allowances the cost of the investment over the useful life of the property. Shorter depreciation schedules, which allow investments to be depreciated quickly, are favored by investors as they allow costs to be recovered more quickly.[10]

Bonus depreciation has also been used to encourage investment in renewables.[11] The Emergency Economic Stabilization Act of 2008 (EESA; P.L. 110-343) provided a 50% first-year bonus depreciation for eligible renewable energy investments.[12] This provision was extended through 2009 under ARRA. Bonus depreciation for renewables expired at the beginning of 2010, before being retroactively extended through the end of the year under the Small Business Jobs Act of 2010 (P.L. 111-240). The Tax Relief, Unemployment Reauthorization, and Job Creation Act of 2010 (P.L. 111-312) further expanded and extended bonus depreciation provisions. Under P.L. 111-312, qualifying renewable energy property placed in service between September 8, 2010, and January 1, 2012, qualifies for 100% bonus depreciation (i.e. the property may be expensed). Qualified property placed in service in 2012 qualifies for 50% bonus depreciation.

Creation of Section 1603 was motivated by challenging economic conditions and reduced availability tax equity, which is the primary vehicle for renewable energy projects to monetize tax incentives.[13] Section 1603 grants allow renewable energy project developers to immediately recover up to 30% of eligible project capital cost expenditures. This serves to, in some cases, reduce the financial risk[14] for projects and provide an alternative tax credit monetization pathway.

Renewable Energy Market Response to Section 1603

Participation in the Section 1603 grant program includes several renewable energy technologies. As of December 6, 2010, grant awards had been made to 1,495 entities with a total grant award value of $5.6 billion. Entities that installed solar electricity projects have received the largest number of grant awards, while wind projects have captured most of the grant dollars (see Figure 1).[15] "Other" projects awarded grants include all other ITC- and PTC-eligible property. A state-level analysis of the Section 1603 program, which indicates the number and value of grant awards made to each state, is included as an Appendix to this report.

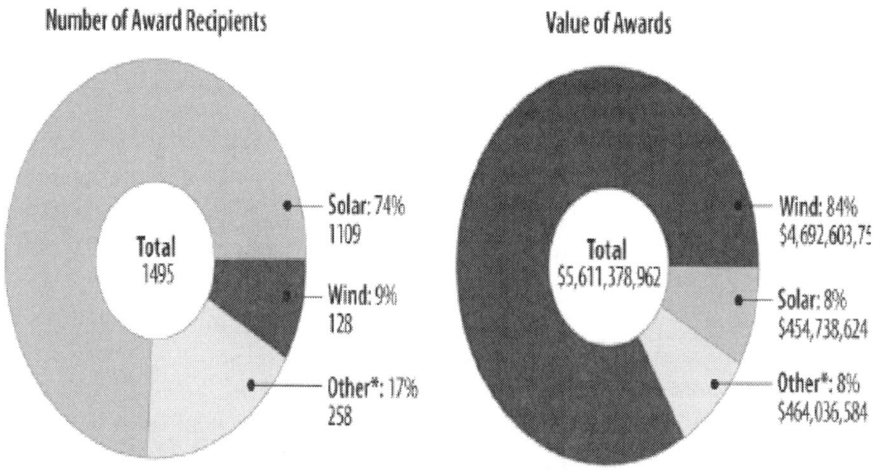

Source: U.S. Department of the Treasury.

Notes: Numbers in this chart are based on Treasury data as of 12/6/2010. Correspondence with U.S. Treasury officials revealed that the total number of projects that had received grant awards, as of 12/2/2010, was 4,215 with a total award value of $5.7 billion. The large discrepancy with project award numbers is a result of some entities receiving multiple awards. The discrepancy in award value is likely a result of a delay in updating the publicly available spreadsheet (used for this analysis) prior to the email correspondence. Not enough information was available to allow for total reconciliation of these discrepancies. * "Other" projects may include data that was mis-categorized.

Figure 1. Section 1603 Cash Grants Awarded To Date.

Market response and impact, along with motivations to apply for a Section 1603 cash grant, differ among the technology types. The following analysis evaluates these parameters for the wind, solar, and other technologies that have elected a cash grant in lieu of tax credits.

Wind

Wind projects have received, by far, the most value (in terms of dollars) from the Section 1603 grant program. Approximately $4.6 billion in grants have been awarded to 128 wind projects. In 2009, the first year the cash grant was in effect, a record 10 Gigawatts (GW) of wind capacity was installed in the United States (see Figure 2). However, it is questionable what portion of 2009 installations were enabled by the 1603 cash grant. Since large wind projects can take 9 to 12 months or more to complete many projects that became operational in the first three quarters of 2009 would have likely been completed without the Section 1603 cash grant. Despite these difficulties, researchers have attempted to quantify the impact of the Section 1603 grant program on the industry. According to one estimate, as much as 2.4 GW of 2009 wind installations were directly enabled by the Section 1603 cash grant program.[16]

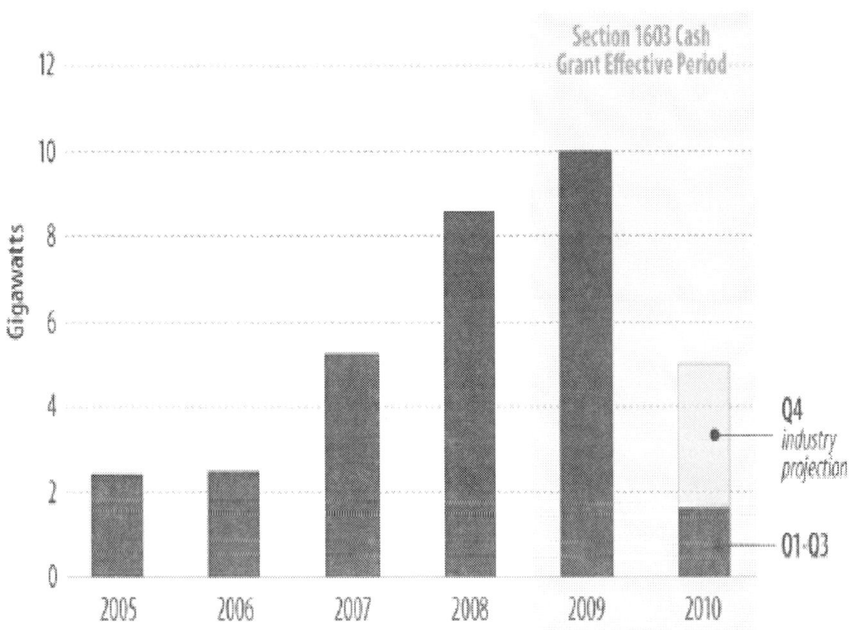

Source: American Wind Energy Association (AWEA) annual and quarterly reports.

Notes: Section 1603 became law in February 2009. Program guidance from the Department of the Treasury was released in July 2009.

Figure 2. U.S. Wind Installations Since 2005.

Wind installations in the first three quarters of 2010 totaled 1.63 GW, a near 70% decrease from 2009 levels.[17] Furthermore, many industry projections estimate that CY2010 large wind installations will be approximately 5 GW, a nearly 50% reduction from 2009 levels. Given these figures, critics of the Section 1603 cash grant may argue that the program had marginal positive impacts on the wind industry. Interpreting such industry statistics should be done cautiously, however, since the Section 1603 cash grant is only one of several factors that influence annual wind capacity additions. Other factors may include:

- **General economic conditions (electricity demand):** Electricity demand is directly correlated to economic conditions, and new generation requirements decrease during periods of economic slow down. AWEA attributes the 2010 wind installation downturn, in part, to "sluggishness in the economy."[18] According to EIA data, in the first eight months of 2010 U.S. electricity demand had rebounded to 2008 levels following a drop of approximately 4.7% in 2009.[19] The potential lag effect of energy demand reductions, and therefore renewable energy procurement, is beyond the scope of this report.
- **Transmission access and availability:** Access to transmission capacity is a critical parameter for wholesale energy projects as it provides a mechanism for matching the producer with consumers. Many high quality wind sites in the upper midwestern United States are limited by the availability of transmission infrastructure to deliver wind-derived energy to demand centers.
- **Natural gas prices:** Purchasers of wholesale power commonly use the cost of natural gas electricity generation as a benchmark for procuring renewable electricity. For example, the California Public Utilities Commission (CPUC) uses the "Market Price Referent," or MPR, as a benchmark for assessing above-market costs of renewable electricity generation and the MPR "can serve to contain the total cost" of renewable energy procurement programs. California's MPR is based on a 500MW Combined Cycle Gas Turbine electric generating facility and includes natural gas price projections.[20] With natural gas prices at relatively low levels, the proxy for renewable energy procurement is low as well, and this condition makes it difficult for project developers to establish Power Purchase Agreements (PPAs) with utility customers that result in an economically viable project. Renewable energy projects

in Virginia, Kentucky, and other states reportedly have been rejected due to lower fossil energy generation costs.[21]

- **Policy clarity:** The American Recovery and Reinvestment Act (ARRA) became law in February 2009 and Treasury guidance for the program was released in July of 2009. Industry participants might argue that financing decisions and commitments were at a near standstill during this period since the financial community was not able to clearly understand the impacts of the Section 1603 cash grant program until details of the program were available. This time lag between policy announcement and policy definition could be a possible contributor to the decline of wind installations in 2010.

A complete analysis of the Section 1603 grant program requires evaluating the value of pre-2009 tax incentives to the benefits associated with receipt of a Section 1603 grant. As discussed above, prior to 2009, the primary tax incentives for investments in wind energy were the PTC and accelerated depreciation.[22] With a PTC, projects with higher capacity factors (percent of generating time per year) would receive the most value from PTC incentives. Furthermore, accelerated depreciation allows a project to maximize depreciation losses in the first five years of operation and this serves to reduce taxable income, and therefore improve project cash flow, during this five-year time period.

Wind and other renewable energy projects, especially in the early years of operation, often do not have taxable income sufficient to maximize the value of the PTC and accelerated depreciation deductions. Therefore some wind projects, prior to Section 1603 availability, would essentially monetize the PTC and depreciation tax benefits, at a discount, through a tax equity investor (a definition and description of tax equity is provided in this paper under the section heading "Section 1603 and the Renewable Energy Tax Equity Market ") and use the cash proceeds to cover capital and installation costs.

With the availability of Section 1603, wind projects can now choose between the PTC, ITC, or cash grant. PTC values are based on the total amount of electrical generation (kilowatt hours) and cash grant values are based on the total amount of eligible installed costs. In order to assess the relative value, to a wind project, of both financial incentives, a present value model was created to provide comparative estimates for each incentive type. This model provides an illustration of the potential benefits associated with a Section 1603 grant for a hypothetical project. For a specific project, a cash flow model would be the ideal method for estimating these incentive values.[23] However, since all projects have unique financial parameters, this present value methodology was used to provide reasonable and relative numbers for a comparative assessment. Absolute numbers for incentive values will vary based on specific project assumptions. The following table illustrates a hypothetical 50MW wind project, finance assumptions, and calculated values for PTC, ITC and accelerated depreciation (i.e. MACRS) incentives.

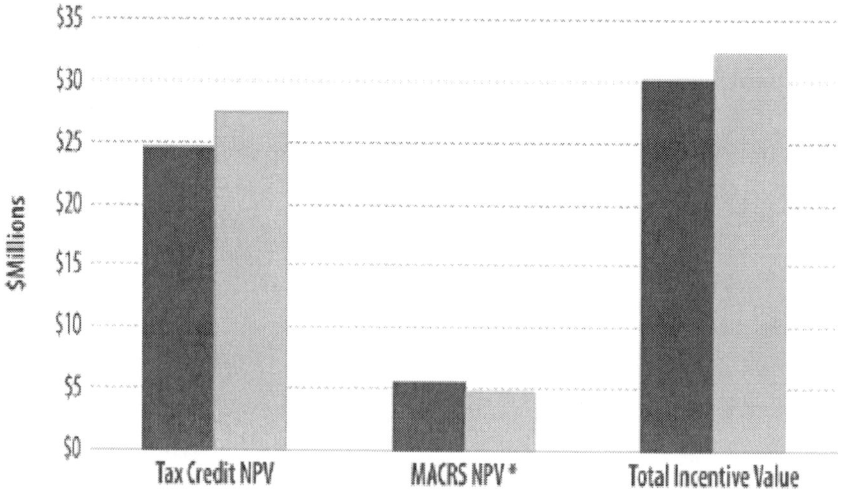

	Tax Credit NPV	MACRS NPV *	Total Incentive Value
PTC	$24.47	$5.64	$30.11
ITC Cash Grant	$27.50	$4.87	$32.37

Source: CRS analysis.

Notes: Project Assumptions: 50MW nameplate capacity; $2/watt installed cost; 35% capacity factor; PTC/kWhr = $0.022 Inflation = 2%; Tax equity required return = 8%; Federal tax rate = 35%; State tax rate = 8%; Effective cash grant rate = 27.5% (this is lower than 30% because not all incurred costs are eligible for the cash grant). NPV = Net Present Value. * MACRS NPV values differ because the depreciation base is reduced by an amount equal to 50% of the ITC cash grant, if the cash grant option is selected; five-year MACRS schedule used to calculate accelerated depreciation value; MACRS value represents the "incremental" value of accelerated depreciation (i.e. the difference between MACRS and 10-year straight-line depreciation).

Figure 3. Example Wind Project: PTC and ITC Cash Grant Incentive Values.

Based on the analysis above, the total value of federal tax incentives (PTC/ITC plus accelerated depreciation), after being discounted based on an assumed tax equity investor required return, is equal to 29% to 32% of the project's total capital cost under both incentive structures. Literature sources cite examples of projects being subsidized by as much as 65% of total project capital costs, although details supporting these estimates are not provided in the references.[24] It is important to understand that the above example is for a specific wind project with pre-defined parameters, is provided as a representative example of calculating federal incentive values, and results will vary based on unique project attributes (i.e., the value of the PTC incentive totally dependent on the project capacity factor).

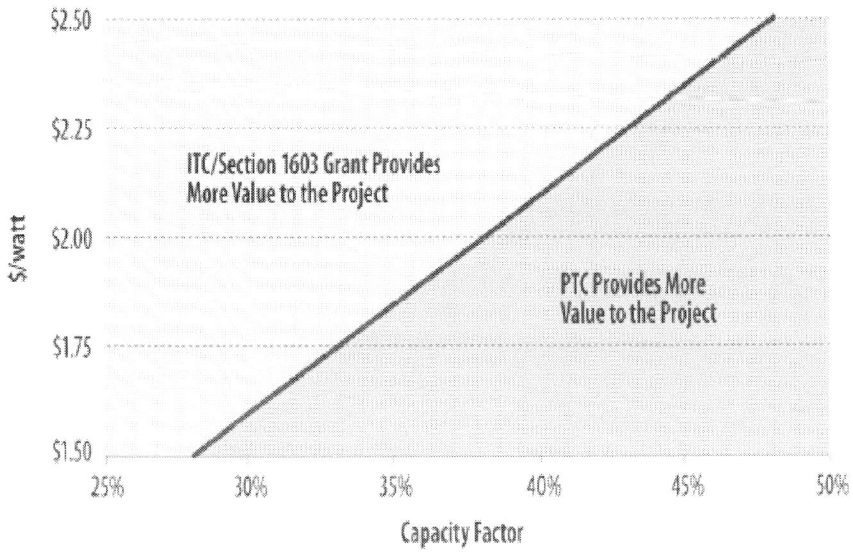

Source: CRS analysis.

Notes: Project assumptions used for this analysis are the same as those noted in Figure 3; PTC and ITC values used to create this figure include the value of MACRS accelerated depreciation; Lawrence Berkeley National Labs (LBNL) performed a similar type of analysis (see Mark Bolinger, Ryan Wiser, and Naim Darghouth, "Preliminary Evaluation of the Impact of the Section 1603 Treasury Grant Program on Renewable Energy Deployment in 2009", Ernest Orlando Lawrence Berkeley National Laboratory, April 2010, page 19.). These results may vary from other estimates as a result of different assumptions used for the calculations.

Figure 4. Incentive Value Sensitivity Analysis.

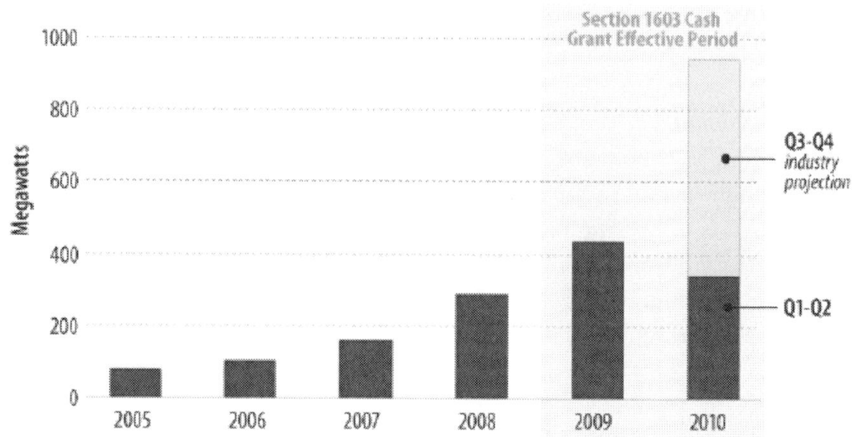

Source: Solar Energy Industries Association.

Notes: Solar installations include all solar market segments: Residential, Non-residential, Utility, and Concentrating Solar Power (CSP).

Figure 5. U.S. Solar Installations Since 2005.

Project capacity factor and installed cost are two primary considerations when evaluating the PTC versus the ITC/Section 1603 cash grant options. Generally, higher capacity factors tend to result in the PTC being the most valuable option, since PTC incentives are based on the amount of annual kilowatt hours produced. However, project capital costs can change the value assessment outcome, since higher capital costs will result in the ITC/Section 1603 cash grant providing more value. The following figure provides a sensitivity analysis for wind projects based on project capacity factor (%) and installed cost ($/watt) metrics.

With the availability of Section 1603 cash grants, wind project developers now have a choice when selecting available federal incentives to support project finance requirements. Depending on the capacity factor and installed cost for a particular project, project developers can quickly determine which incentive structure offers the most value based on these two project parameters.[25] Depending on the capital cost and capacity factor for a specific project, the developer will likely choose the incentive approach that yields the most value. However, the choice of incentives is predicated on the availability of tax equity investors to monetize the PTC tax benefits for wind projects.

One possible effect of the ITC/1603 cash grant option is the development of "marginal" (i.e., low capacity factor) project sites that would not be economically feasible under a PTC incentive structure.[26]

Solar Electricity

The solar electricity marketplace (photovoltaic, solar thermal, etc.) has experienced consistent upward trajectory since 2005 (see Figure 5) and 2010 is projected to be a record year for the industry, with total installed capacity estimates at approximately 1GW (nearly 120% growth over 2009 solar installations).[27] It should be noted, however, that the solar electricity market is currently much smaller than the wind market. While year-on-year growth for solar has been substantial, projected 2010 solar capacity installations, in terms of megawatts, are approximately 15% to 20% of 2010 wind installations.

Solar electricity projects have received the largest number of Section 1603 cash grant awards. The Department of the Treasury has awarded grants to 1,109 entities that represent more than 3,700 individual solar projects.[28] The majority of these projects consist of small installations as evidenced by the distribution of grant award values (see Figure 6).

Compared to the wind marketplace, solar electricity projects are impacted by the Section 1603 cash grant program in different ways. Unlike wind, solar electricity projects do not have a production tax credit (PTC) option to consider when evaluating federal incentive options (solar electricity projects are, however, eligible for MACRS accelerated depreciation). Instead, solar electricity projects can choose either (1) a 30% investment tax credit (ITC), or (2) a Section 1603 cash grant in lieu of the ITC equal to 30% of eligible project capital costs. Option 1 (30% ITC) can sometimes be challenging for solar electricity projects to use, because it requires each project to self-shelter the ITC, using 100% of the tax credit to offset tax liability. Investment tax credits have the most value if they are used to offset income in the current year. If a taxpayer cannot claim all the full value of tax credits in the current years, taxpayers are able to carry forward unused tax credits for up to 20 years.[29] The ITC also has a one year carryback period.[30] For many business entities, especially small businesses with limited revenue and income, having a taxable income base large enough to take full advantage of the tax credit and accelerated depreciation benefits can be difficult. Based on the number of small grant awards issued to date, Section 1603 cash grants appear to have reduced this difficulty by encouraging thousands of small solar electricity project installations.

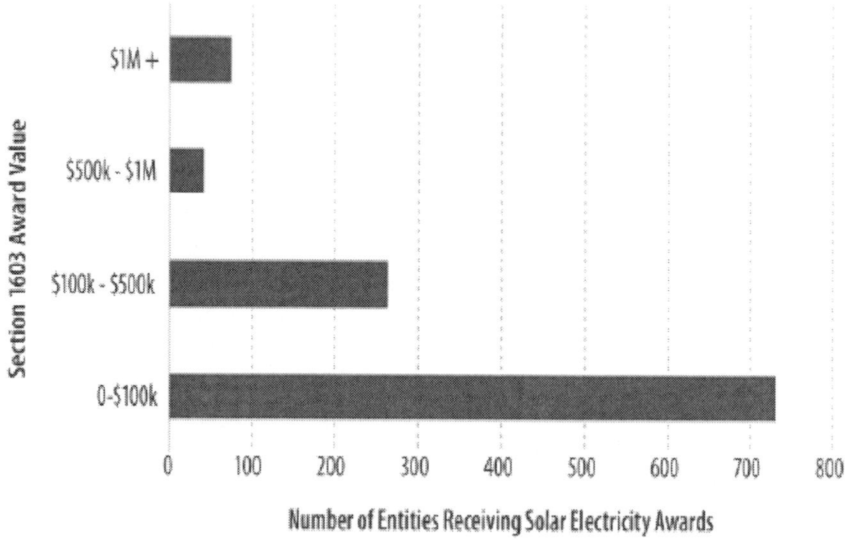

Source: U.S. Department of the Treasury.

Figure 6. Solar Electricity Section 1603 Grant Award Values.

Prior to the availability of Section 1603 cash grants, solar electricity projects that did not have an adequate tax liability to effectively use available tax credits could make arrangements for ITC and MACRS accelerated depreciation monetization with a tax equity investor. Solar electricity projects could, in effect, monetize their ITC and depreciation incentives through a tax equity investor (a definition and description of tax equity is provided in this paper under the section heading "Section 1603 and the Renewable Energy Tax Equity Market ") at a discount and receive cash proceeds towards project capital and installation costs. The ability of individual small solar projects to monetize tax incentives is questionable, though, as the amount of due diligence required for a tax equity investor to structure a transaction may negate the tax benefits offered by small projects.

With the availability of the Section 1603 cash grant, solar electricity projects can choose (1) the conventional ITC option by either monetizing the ITC and accelerated deprecation or selfsheltering the tax benefits, or (2) projects can elect to receive a 30% Section 1603 cash grant while still benefitting from accelerated depreciation (MACRS depreciation deductions can be either self-absorbed or third party monetized). It is also important to point out that Section 1603 cash grants are typically received within 60 days of filing an application. While the "value" of the 30% ITC and the Section 1603 cash grant, in theory, should be equal, in practice the cash grant will likely be perceived as being more valuable than the ITC for the following reasons:

- **No transaction costs:** Realization of investment tax credits may require the project to monetize incentives at a discount to a tax equity investor and also demands personnel time and professional fees to complete the investment transaction. Section 1603 cash grants only require an application to be filed with the U.S. Department of the Treasury with payments typically received within 60 days.
- **No competition for tax equity:** A limited number of tax equity investors are available in the marketplace and projects that decide to monetize the ITC must compete with other projects for tax equity investment deals.
- **Time value of money:** Section 1603 cash grants are paid to projects within 60 days after application submission, while the ITC is claimed when tax returns are filed. For projects unable to fully claim tax credits in the first year, tax credits are carried forward to offset future tax liability. While the timing difference for the two options may not differ dramatically, the cash grant does provide a higher degree of certainty, with regards to cash flow, and flexibility to the project owner.
- **Taxable income risk:** Projects that elect the ITC have risk exposure associated with the adequacy of taxable income to absorb the full value of ITC incentives. A cash grant eliminates this risk.

An emerging segment of the U.S. solar photovoltaic (PV) market includes what is known as "Utility Scale" projects, which are typically multi-megawatt, centralized, grid-connected solar generation facilities. Approximately 214 MW-dc (Megawatts – direct current) of operating utility-scale PV projects exist in the U.S. today, while more than 5,362 MW of utility-scale solar projects are under contract (meaning they have a signed power purchase agreement, or PPA, with a third-party to purchase the electricity generated) and an additional 10,000 Megawatts have been announced (see Figure 7).[31]

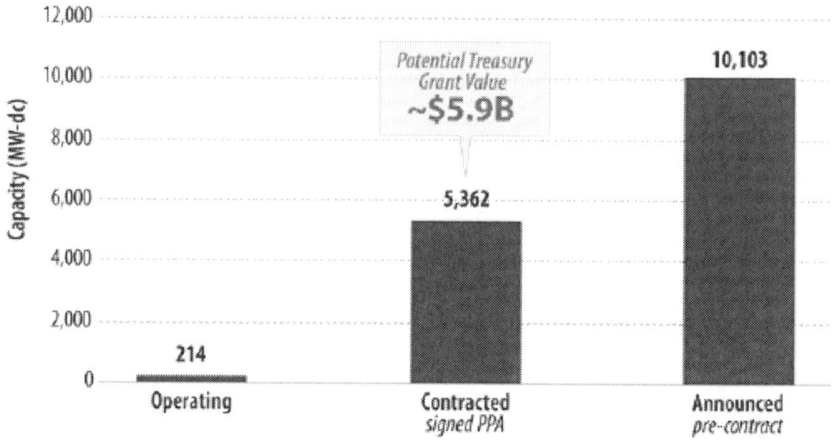

Source: Shayle Kann, "The Future of the Utility Scale PV in the U.S.", *Greentech Media*, December 1, 2010.

Notes: "Potential Treasury Grant Value" of $5.9 billion is calculated based on the following two assumptions: (1) Total capital cost = $4/watt; (2) Effective treasury grant for these projects = 27.5% (this is less than 30% since not all capital costs are eligible for the treasury grant).

Figure 7. Projected Utility Scale PV Capacity Development Status.

While it is unlikely that all contracted and announced projects will be constructed (some of these projects may encounter permitting, transmission access, financing, and other hurdles that could prevent their development), this analysis illustrates the maximum value of cash grants that might be awarded to utility-scale PV projects, assuming that all projects are eligible for, and elect, a Section 1603 cash grant. Additional solar technologies, including CSP (Concentrating Solar Power) and CPV (Concentrating Photovoltaic), may also pursue Section 1603 cash grants.

"Other" Technologies

A third category of Section 1603 cash grant awards is "other" technologies, which includes all eligible renewable technologies except wind and solar. This category represents approximately 8% of total grant award dollars with an aggregate value of approximately $461 million. The figure below provides a breakdown of grant value received by the "other" technologies (see Figure 8).

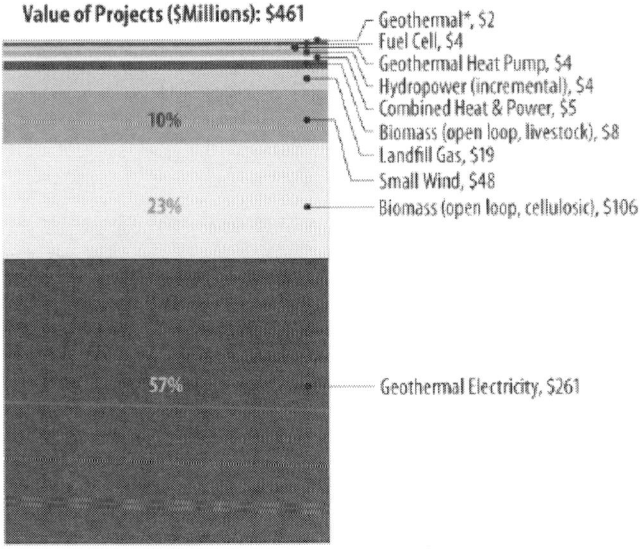

Source: U.S. Department of the Treasury.

Notes: *Award information provided on the Treasury website includes two projects categorized as "Geothermal" and it is not clear how these two projects are distinctly different than "Geothermal Electricity" and "Geothermal Heat Pump" projects.

The total value of "small wind" projects is likely to be much lower due to an apparent mis-categorization of a wind project that received a $43 million cash grant.

"Microturbine" and "Marine" projects have not received any Treasury grant awards and, therefore, are not listed in the figure.

Figure 8. "Other" Technologies: Section 1603 Grant Award Analysis.

The number of projects that have received awards in the "other technologies" category is relatively small compared to wind and solar electricity. Possible reasons for this may include (1) the financial community has a certain comfort level with wind and solar electricity project deals and may be wrestling with how best to structure project finance for "other" technologies; (2) securing attractive off-take agreements (i.e., Power Purchase Agreements) may be challenging for this set of technologies; (3) technology maturity and performance risks for some of the "other" technologies; (4) for some of the "other" technologies, longer development timelines introduce a degree of financial risk during the development phase of the project; and since tax equity, bridge loans, and project debt are typically committed during the latter stages of development/ construction, some projects may not have the financial capability to fund longer development timelines.

The "other" technologies grant award category is dominated by three technology types: (1) Geothermal electricity; (2) Biomass (open loop, cellulosic); and (3) Small wind. Since the "other" technologies represent a relatively small portion of total grant awards, a rigorous assessment of each technology was not conducted in preparation of this report. However, the following discussion provides a condensed analysis of geothermal electricity, biomass, and small wind projects that have received Section 1603 cash grants.

Geothermal Electricity

Within the "other" technologies category, geothermal electricity projects have received 57% of grant award values representing approximately $261 million. Six geothermal projects have received Section 1603 cash grants to date, although five of these projects account for 99.9% of the $261 million of grant awards.[32] The five largest geothermal electricity projects were installed in three states: Nevada (3), Utah (1), and California (1). Installed nameplate capacity for each of these projects range from 18MW to 100MW.[33] Geothermal energy properties qualify for PTC, ITC, or cash grant incentives. Selection of the optimal incentive for geothermal energy property may be similar to the process described above for wind energy property.

Biomass (Open-Loop, Cellulosic)[34]

To date, cellulosic biomass electricity generation projects have received approximately $106 million of Section 1603 cash grant awards. Twelve cellulosic biomass electricity projects, in eleven different states, have received awards to date. Biomass may include "a large range of feedstocks ... from woody and herbaceous biomass to agricultural residues."[35] Much like wind, biomass energy properties have the option of choosing PTC, ITC, or cash grant incentives. Therefore, selection of the best incentive for a particular project may be similar to the process described above for wind energy property.

Small Wind

Small wind facilities are distinguished from other wind projects based on the size of the turbine used to produce energy. Specifically, small wind projects include turbines with a nameplate capacity that does not exceed 100 kW.[36] Small wind Section 1603 cash grant awards include 81 projects with a total grant award value of $48 million. Most of the small wind project awards received range in value between $30,000 and $100,000.[37] Small wind incentive options, which include either an ITC or a cash grant, are similar to those available to solar electricity energy property. The above discussion about solar electricity may provide some insight into why small wind energy properties are motivated to elect the cash grant option.

SECTION 1603 AND THE RENEWABLE ENERGY TAX EQUITY MARKET

The Section 1603 cash grant program was created as an alternative mechanism for renewable energy projects to realize federal tax incentives. Specifically, the program was created to address the limited availability of tax equity investor funds resulting from challenging U.S. economic conditions. U.S. Department of the Treasury guidance for the Section 1603 program states that "the Section 1603 program will temporarily fill the gap created by the diminished investor demand for tax credits."[38] In order to better understand the mechanics and function of tax equity investors, the following is a brief overview and analysis of the tax equity marketplace to include an assessment of available tax equity to support development of renewable energy projects. Analyzing the tax equity market is challenging since there is no official barometer or tracking mechanism available to assess the history and status of the tax equity marketplace. A survey of multiple industry sources, including statements from tax equity market participants and experts, was used to assess the history and current state of the U.S. renewable energy tax equity market.

What is "Tax Equity"?

Tax equity is a hybrid (debt/equity) type of investment that has a preferred position, over the project sponsor, for the project cash flows and tax benefits. For a renewable energy project, tax equity is a form of investment that allows the investor to use tax benefits such as losses created by accelerated depreciation and production/investment tax credits that would otherwise not be effectively used by the project itself (taxable income for renewable energy projects is typically not large enough to absorb these tax incentives). In essence, a renewable energy project can monetize (monetization is simply the conversion of future tax benefits into cash), at a discount, its federal tax incentives (PTC, ITC and accelerated depreciation) through an investor for cash that can be used to pay for a portion of project capital and installation costs. Much like traditional equity, "tax equity" takes risk on the project performing and operating as expected, but like debt "tax equity" has preferential financial treatment, with regards to cash flows, and typically requires a moderate rate of return (less than traditional equity; equal to or greater than debt rates).

Who are Renewable Energy Tax Equity Investors?

To date, most renewable energy tax equity investments have been made by banks, financial companies, investment banks, and insurance companies. These entities either have a taxable base large enough to realize the value of federal tax incentives for renewable energy projects or they aggregate a group of clients/third parties that can capture the incentive value. Specific firms with activity in the tax equity marketplace include JPMorgan, GE Capital, Wells Fargo, Morgan Stanley, and Credit Suisse, among others. These institutions are motivated to provide financing to renewable energy projects, since tax equity investments provide an opportunity to realize an attractive financial return on investment, partially by reducing total tax liabilities as well as through cash generated from the projects. Risks associated with these returns include subpar operational performance of the projects (especially wind, which is incented to maximize electricity production) and an inability of the investment entity to absorb tax incentives over several years (typically 5 to 10 years).

Renewable Energy Tax Equity Market Analysis

Since there are no official sources or statistics for tax equity, performing a comprehensive analysis of the renewable energy tax equity market, its size, and how it has changed, in light of recent economic conditions, is difficult. However, a disparate set of information and third-party analysis of the tax equity market was aggregated and compared to provide some insight and analysis into this relatively obscure project finance investment type. Table 2 provides a summary of the renewable energy tax equity marketplace based on three different industry sources.

Table 2. Renewable Energy Tax Equity Market Analysis

	Chadbourne & Parke Reports				Bloomberg-NEF				USPREF			
	2007	2008	2009	2010	2007	2008	2009	2010	2007	2008	2009	2010
# of Tax Equity Investors	14	9-18[a]	4	5-15[a]	16	n/a	5	n/a	20	13	11	16
Tax Equity Yields (%)	n/a	7.5-7.9	8.1-8.4	8-8.5	6.3	9	8->10	8.1	n/a	n/a	n/a	n/a
Total Amount of Tax Equity ($B)	5[b]	2.6[b]	1.1[b]	4.3[a]	n/a	n/a	n/a	n/a	6.1	3.4	1.3	3

Sources: Chadbourne & Parke Project Finance NewsWire reports from the following months: November 2008, January 2009, February 2010, April 2010, and November 2010. Bloomberg New Energy Finance, “Cash is king: shortcomings of US tax credits in subsidising renewables” January 20, 2010. Bloomberg New Energy Finance, “A bang or a whimper? Implications of the US Treasury cash grant expiration,” August 2010. U.S. Partnership for Renewable Energy Finance, “Prospective 2010-2012 Tax Equity Market Observations,” July 2010.

Notes:

Discrepancies associated with the “Total Amount of Tax Equity ($B)” numbers in the chart are partly due to some sources reporting these numbers as “raised and committed” versus “actually funded.”

2010 Tax equity values ($B) include deals that provided “bridge” funding for the cash grant.

NEF = New Energy Finance

USPREF = U.S. Partnership for Renewable Energy Finance

[a] Ranges in Chadbourne Parke reports are based on expert comments at different times of the year.

[b] Tax equity numbers for years 2007-2010 is for Wind only; 2010 estimate is for the first three quarters of 2010; 2008 and 2009 numbers are not from the literature, but rather were provided as unofficial estimates from an industry source.

Table 2 provides an overview of the renewable energy tax-equity marketplace and changes observed since 2007. Some key takeaways from this tax equity market analysis include:

- Calendar year 2007 experienced the largest amount of renewable energy tax equity investment. An estimated $6.1 billion of "tax equity" went to various projects with approximately $5 billion dedicated to wind.
- In 2008, both the total volume of deals and the number of tax equity investors declined. Poor economic conditions are one possible explanation for this decline.
- Tax equity yields (the financial return to the tax equity investor) have fluctuated between 7.5% and 10% since 2007 and appear to have settled in the 8%-8.5% range. This fluctuation may be related to the volatility in the number of tax equity investors during the period (2007-2010).
- The number of tax equity investors and the amount of tax equity continued to decline in 2009. While some may argue that this is evidence of inadequate tax equity, others may argue that since Treasury guidance for the Section 1603 program was not released until August of 2009, the financial community was waiting until Treasury guidance was available before making any funding commitments.

One additional observation in the above summary chart is that the tax equity market appears to be recovering from its 2009 lows. Data for 2010 renewable energy tax equity market activity indicate that the number of investors is near record levels and transaction dollar volumes are increasing. On the other hand, comments from industry experts in the literature indicate that some tax equity investors currently (2010) in the market "could not have done a production tax credit or investment tax deal in 2007 and could not do one today."[39] In other words, according to this industry expert, the Section 1603 grant program has enabled participation by some tax equity investors that otherwise would not have made investments in renewable energy projects. According to industry expert commentary in the literature, Section 1603 cash grants allow tax equity investors to quickly recover a large portion of their investment, which in turn reduces the risk profile for tax equity investments.[40] This in turn reduces the required tax capacity for the investor and the investment recovery horizon is reduced from 10 years (typical under the traditional tax incentive structure) to 5 years or less (typical for cash grant tax equity deals).

In essence, availability of Section 1603 cash grants has changed the nature of tax equity investment deals and has also resulted in the emergence of new financial structures for renewable energy projects. Renewable energy investors now have multiple options to consider when financing projects.[41]

One alternative approach for evaluating the tax equity marketplace at a macro level is an assessment of all U.S. corporate profits. Figure 9 shows corporate profits from 2001 to 2009. This figure represents total corporate profits (not profitability for firms that typically make renewable energy investments) and it provides a general sense of profitability trends in the United States since 2001. Since corporate profits are an indicator of taxable income, this macro-level information provides some degree of insight into tax credit capacity trends.

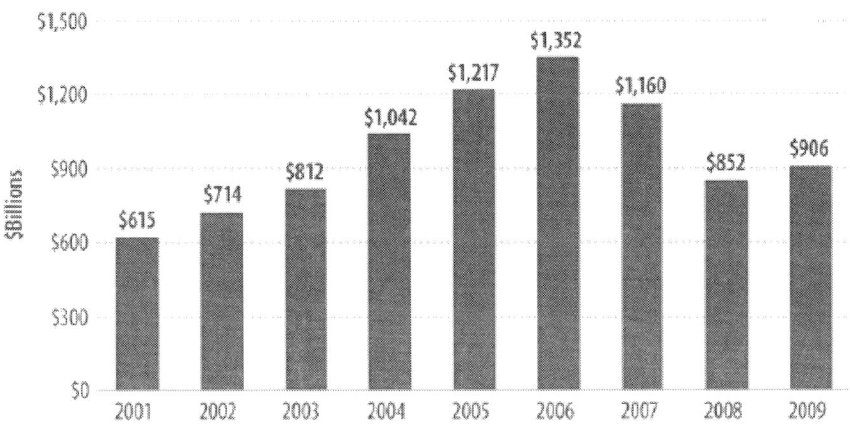

Source: U.S. Department of Commerce - Bureau of Economic Analysis.

Notes: Domestic Industries - Corporate profits with inventory valuation and capital consumption adjustments.

Figure 9. U.S. Corporate Profits: 2001-2009.

This data set illustrates that domestic U.S. corporate profitability has decreased from a recent peak in 2006. With regards to renewable energy tax equity investments, a specific analysis of the profitability of firms that typically make renewable energy tax equity investments may provide a more detailed estimate of tax equity capacity for these types of projects. Nevertheless, this macro view of corporate profitability does provide an indication of profitability, and therefore tax equity capacity, trends in the

United States. One could argue that total domestic corporate profits represent a theoretical maximum limit for tax equity investments. However, a more realistic practical limit will likely be much lower since some companies may not have the capability, expertise, or comfort level needed to make investments in renewable energy projects.

IMPACTS OF THE 1603 GRANT PROGRAM

The following sections provide additional analysis of the impacts of the Section 1603 grant program. The Section 1603 grant program has been popular with the renewable energy sector. Proponents of the program suggest that the added incentive is necessary to continue to promote renewable energy. The Section 1603 grant program, however, results in revenue losses that are greater than the revenue losses associated with the previously available tax incentives. Given the country's large budget deficits, there may be questions of whether further extensions of this program are worth the budgetary cost.

Economic Considerations of Section 1603 Implementation

From an economic perspective, energy tax provisions can be evaluated by asking two questions. First, does the provision address an energy market failure? Second, if the provision successfully addresses an energy market failure, is the provision the most economically efficient solution. The Section 1603 grants do address a perceived failure in energy markets, specifically use of fossil energy resources, which may have environmental and potentially energy security effects. As is often the case with energy tax incentives, however, the grants may not be the most economically efficient tool available for addressing this market failure.

A number of potential market failures may be used to rationalize government intervention in energy markets.[42] For example, in the electricity production market, externalities may lead to a market failure.[43] Electricity produced from fossil energy sources may have unpriced negative environmental consequences not taken into account when energy production and consumption choices are made.[44] If externalities are present, markets fail to establish energy prices equal to the social marginal cost of supply. The result is a system where cost and/or price signals are inaccurate, such that the socially optimal level of output, or allocative efficiency, is not achieved.

Economic theory suggests that a tax be imposed on activities associated with external costs, while activities associated with external benefits be subsidized—in order to equate the social and private marginal costs. These taxes and/or subsidies would, theoretically, result in a more efficient allocation of resources.

Oftentimes in energy markets, instead of taxing the activity generating the negative externality, policymakers have instead chosen to subsidize the non-polluting alternative. The Section 1603 grants are an example of such a subsidy. Section 1603 grants reduce the price of investment in renewable energy, encouraging development of additional renewable capacity. Subsidizing the alternative activity, rather than taxing the polluting activity directly, is an economically inefficient policy option.

Subsidies that encourage the use of alternative or renewable energy sources, as opposed to a direct tax on polluting energy sources, are economically inefficient for two reasons. First, subsidies reduce the average cost of energy. As the average cost of energy falls, the quantity demanded of energy increases, countering energy conservation initiatives. Second, paying for subsidies requires the government to raise funds elsewhere. If these funds are raised through a distortionary tax, such as a tax on labor, such subsidies are economically inefficient.[45] This concern is not unique to the Section 1603 grant program, as it also applies to the use of the PTC and ITC to promote investment in renewable energy.

Section 1603 grants may also be economically inefficient to the extent there are inframarginal beneficiaries. Inframarginal beneficiaries are projects that would have been installed without a grant, but receive a grant nonetheless.[46] The other type of recipients, marginal beneficiaries, are projects that were able to be installed only because the grant was available. The grant provides a windfall benefit to inframarginal beneficiaries, without resulting in additional installations of renewable energy generation capacity. The most cost-efficient tax incentives are those that have large marginal impacts. In other words, cost-efficient programs are those that lead to greater participation in the subsidized behavior, without simply rewarding those already engaged in the subsidized behavior. Again, this concern is not unique to the Section 1603 grant, as it applies to energy tax incentives broadly. The analysis presented in the "Wind" section above highlights how the additional incentive provided by the Section 1603 grant option might attract additional installations. As tax equity markets recover, increasing projects' ability to move forward without the grant, the number of inframarginal beneficiaries may increase, thereby decreasing the overall economic efficiency of the program.

Compared to tax incentives, the Section 1603 grants, or grants generally, may be a more efficient subsidy. From the perspective of the federal government, the efficiency of a subsidy can be evaluated based on funding flowing toward the subsidized activity relative to federal revenue losses.[47] Since many recipients of renewable energy tax incentives turn to tax-equity markets to monetize such credits at a discount, it may be more efficient for the government to provide grants directly rather than use the tax code as a vehicle for subsidization. Further, this would eliminate the need for tax-equity markets broadly, potentially leveling the playing field for smaller projects with less access to tax equity. Policymakers may be less inclined to use a direct grant approach, relative to a tax incentive or grant tied to a tax incentive option, as direct grants involve government outlays and may be subject to greater budget scrutiny and the annual appropriations process.

Cost of Section 1603 to the Federal Government[48]

There are various ways to evaluate the cost of the Section 1603 grant program. It is important to remember that the Section 1603 grant program is sequentially related to the PTC and ITC. First, PTC eligible taxpayers may elect to receive an ITC instead of the PTC. Next, taxpayers can request a grant from the Treasury in lieu of the ITC. The option to claim the ITC in lieu of the PTC is expected to increase revenue losses associated with the ITC. As ITC-revenue losses increase, revenue losses associated with the PTC may decrease as fewer PTC claims are made in the out years.[49] Further, since the ITC is awarded at the time of investment, while the PTC is awarded for production over a 10-year period, shifting to the ITC will shift revenue losses forward as expenditures are made in the near term. Finally, as renewable energy taxpayers elect to receive a direct payment from the Treasury in lieu of tax credits, the Section 1603 grant program will result in federal outlays.

Table 3 presents the Joint Committee on Taxation's estimated revenue effects under ARRA. When the Section 1603 Grant program was established, the JCT estimated that the program would result in 10-year revenue losses of $218 million as taxpayers shifted from the PTC to the ITC, 10-year revenue losses of $158 million as grants were paid out, and $153 million in 10-year revenue gains due to reduced ITC and PTC claims in the future. As is noted below, subsequent JCT revenue scores suggest that the Section 1603 grant program is more expensive than was initially anticipated.

Table 3. Estimated Cost of 1603 Grants Under ARRA: JCT millions of dollars

	2009	2010	2011	2012	2013	2009-2013	2009-2018
Elect ITC in lieu of PTC	-96	-131	-29	16	11	-230	-218
Grant in lieu of tax credit: outlay effects	-30	-88	-40			-158	-158
Grant in lieu of tax credit: revenue effects	3	11	22	41	41	118	153

The Department of Treasury's estimated outlays associated with Section 1603 are reported in the President's FY2011 Budget (see Table 4).[50] The Treasury estimates that the Section 1603 grant program will result in outlays of $15.4 billion through 2015. To date, program outlays have exceeded the Treasury's expectations. In 2009, $1.6 billion in grants were awarded. In 2010, grant awards were nearly $4.2 billion.

Table 4. Outlays Under the 1603 Grant Program millions of dollars

	2009	2010	2011	2012	2013	2014	2015
Treasury's estimated outlays on grants in lieu of tax credits	1,050	3,090	4,460	4,240	2,360	230	30
Outlays on actual grant awards made	1,643	4,189					

Source: Department of the Treasury, President's FY2011 Budget and Section 1603 list of awards, available at http://www.treasury.gov/initiatives/recovery

Given the structure of the Treasury grant program, it is expected that the actual cost of the program will be less than the outlays under the program. Since the grant is designed to replace tax credits, the cost of the grant will be (at least partially) offset by reduced future tax expenditures.

In other words, projects receiving a grant in lieu of the PTC receive funds from the government when the project is placed in service. However, since that project will not be claiming the PTC in the out years, paying the grant up front reduces PTC payouts in the future. Tax expenditure estimates prepared by both the JCT and Treasury, however, do not indicate decreased PTC revenue losses in the future. PTC revenue losses increase over time as the credit is adjusted for inflation and the amount of PTC-eligible production increases. That the PTC does not increase in the out years suggests that the dampening effect of the Section 1603 grant on PTC revenue losses do not exceed these other factors.

After having some experience with the Section 1603 grant program, the JCT revised the program's estimated tax expenditures.[51] The one-year extension of the Section 1603 Grant program enacted under P.L. 111-312 is associated with nearly $3 billion in federal revenue loss (see Table 5).[52] The cost of the extension alone exceeds the estimated annual revenue losses associated with the PTC and ITC. For FY2010, the tax expenditure estimate for the PTC was $1.4 billion with the ITC costing less than $50 million.[53]

Table 5. Federal Revenue Loss of One-Year Extension of Section 1603 Grant millions of dollars

	2011	2012	2013	2014	2015	2011-2015
One-year extension of the Section 1603 grants in lieu of tax credits	1,941	1,045	—	—	—	2,987

Source: Joint Committee on Taxation, JCX-54-10.

Notes: Total estimate for the 2011-2015 period may not equal the sum of all years due to rounding.

Job Creation from Section 1603

The potential for job creation has become a key factor in evaluating renewable energy investment incentives and programs. Despite being an issue of importance, quantifying and measuring green job creation and growth has been difficult. First, the term "green jobs" is not precisely defined. Second, authoritative data compiling jobs in the green economy do not exist.[54] Nonetheless, some estimates have been made regarding the jobs impact of the

Section 1603 grant program.[55] While estimates have been made, this is not to say that such estimates have successfully overcome the challenges associated with estimating "green job" creation. Thus, it is recommended that any job creation estimate be viewed with skepticism.

Focusing on the wind sector, Bolinger et al. (2010) provide an early estimate of job creation under the Section 1603 grant program.[56] Counting only jobs resulting from projects that would not have been viable without the grant, Bolinger et al. estimate that the Section 1603 grant was responsible for generating 51,600 short-term jobs and 3,860 long-term jobs in the wind industry through March 1, 2010.[57] The American Wind Energy Association (AWEA) claimed that failing to extend the Section 1603 grant for one year, through 2011, would have put 15,000 wind energy jobs in jeopardy.[58] The Solar Energy Industries Association claims that the Section 1603 induced approximately 8,500 jobs in the solar industry through November 22, 2010.[59] Of these, an estimated 2,600 were installation or construction jobs, approximately 5,800 were in solar manufacturing, while approximately 60 jobs were related to operations and maintenance.[60]

To put these figures in perspective, it is helpful to compare estimates of job creation due to the Section 1603 grant provision to estimated job creation under ARRA broadly. The Congressional Budget Office (CBO) has released a series of reports analyzing the job creation and economic stimulus effects of ARRA.[61] The most recent report, covering the 2nd quarter of 2010, estimates that ARRA stimulus funds increased the number of people employed by 1.4 to 3.3 million, lowering the unemployment rate by 0.7 to 1.8 percentage points.[62]

Potential Policy Options Related to Section 1603

When the Section 1603 grants were enacted under ARRA, the temporary grant program established a construction start deadline of December 31, 2010. The Tax Relief, Unemployment Reauthorization, and Job Creation Act of 2010 (P.L. 111-312) provided a one year extension, allowing projects placed in service during 2011 as well as those under construction by the end of the year to qualify for the grant option. The grant option remains popular with renewables investors, and is likely to remain popular through 2011. Given the financial benefits associated with the grant option, the renewable energy sector is likely to request an extension beyond 2011.

In evaluating whether the grant should be further extended, Congress may want to consider a number of factors. First, if Congress decides that subsidizing investment in renewables remains a policy objective, policymakers might evaluate whether the grant is the most economically efficient mechanism for providing subsidization. Second, Congress may want to consider the policy rationale going forward. The grant was initially enacted to address weakness in the tax equity market. As the tax equity market recovers, should the grant program be retained? In practice, the grant program may have increased the use of renewable energy, addressing carbon mitigation and greenhouse-gas reduction policies. A question for Congress is whether these benefits, and the benefits associated with increased investments in renewable energy and potential job creation and economic stimulus, are worth the budgetary cost and overall contribution to federal deficits.

There are a number of options available to Congress regarding the Section 1603 grant program. The current construction start deadline of December 31, 2011, means that, without action, the grants effectively expire at the end of the year. Alternatively, Congress may choose to extend the grants. Extending the grant program could involve extending the construction start date, extending the placed-in-service deadline, or other modifications. Finally, there have been a number of benefits associated with the Section 1603 grants in lieu of tax credits that could be achieved through modifications to the existing PTC and ITC. These potential policy options are discussed in greater detail below.

Allow the Section 1603 Grant Program to Expire

Absent Congressional action, projects beginning construction after December 31, 2011, will not be able to receive the Section 1603 grants in lieu of tax credits. These projects will, however, continue to be eligible for the renewable energy PTC or ITC (the expiration date of these incentives varies by technology). The Section 1603 grant in lieu of tax credit program was established to address limited tax-equity availability following the financial crisis. Since the goal of the program was to prevent a decline in installed renewable energy capacity for projects that typically would have relied on tax equity, if tax-equity markets have recovered the grants may no longer serve their initial purpose.

While the status of tax-equity markets is one consideration, there are other factors that policymakers may want to consider. It has been noted that the Section 1603 grants in lieu of tax credits have provided a number of benefits to renewable energy projects. Benefits from the grant, such as full AMT exemption, relief from PTC "haircuts," and the exemption from passive credit limitations, in some cases may be worth more than the face value of the grant relative to the PTC (discussed in greater detail below).[63] Further, these benefits are available to all renewable energy projects, not just those who would have sought to establish a tax-equity relationship to monetize tax credits in the absence of the 1603 grant program. These benefits, among others, are discussed in greater detail below in the context of possible modifications to the PTC and ITC that would result in benefits similar to those afforded by the grant.

Extend the Section 1603 Grant Program

One option for Congress would be to again extend the Section 1603 grant program on a temporary basis. Temporary extensions are attractive for budgetary reasons, and allow for policy flexibility in the future. Temporary extensions, however, may not be attractive from the industry perspective as potential renewable energy project investors and developers face uncertainty with respect to future financing options.

Several issues arise in considering an extension of the Section 1603 grant program. One important issue is the placed-in-service deadline for wind to qualify for the PTC. Under current law, PTC-qualifying wind facilities must be placed in service by the end of 2012. For other PTCeligible technologies, the placed-in-service deadline is the end of 2013 (see Table 1). Given the current status of the PTC, extending the Section 1603 grant option beyond 2012 would decouple the PTC placed-in-service deadline from that of the grant. If the PTC is no longer available for wind, any grant that continues to be available would no longer be "in lieu" of tax credits.

Historically, PTC expirations have been associated with subsequent declines in installed wind capacity.[64] The importance of long-term incentives for the renewable energy industry was recognized by Congress when enacting ARRA. Under ARRA, the PTC was extended for three years (previous extensions tended to be for one or two years). Thus, as the current Section 1603 grant deadline is only one year shy of the placed-in-service deadline for wind, Congress may wish to consider a more holistic reevaluation of existing tax incentives for renewables, rather than focusing specifically on the Section 1603 grant provision.

Modify Existing Tax Incentives to Enhance Benefits

Changes to the PTC and ITC

Certain features of the Section 1603 grant have provided a number of indirect benefits for renewable energy projects and investors. These indirect benefits are summarized in the shaded text-box below. The first option Congress may want to consider is modifying existing tax incentives to allow for some of the indirect benefits that are afforded by the Section 1603 grant. While a full discussion of the potential effects associated with these policy changes is beyond the scope of this report, some key points are noted below.

Generally, there are more limitations on the PTC than there are on ITC-eligible investments or entities claiming the Section 1603 grant in lieu of tax credits. First, PTC projects cannot use the PTC to offset tax liability under the AMT after the first four years of production. Second, the value of the PTC may be reduced for projects receiving grants or other federally subsidized financing. Finally, PTC-eligible projects must be owner-operated, and must sell their electricity to an unrelated party. Relaxing these restrictions associated with the PTC would increase the value of the credit for some taxpayers while increasing the types of taxpayers and entities that would qualify for the PTC.

Indirect Benefits of the Section 1603 Grants[65]

AMT Relief

Taxpayers claiming the PTC are exempt from the alternative minimum tax (AMT) for the first four years (typically, taxpayers are eligible to receive the PTC for the first 10 years after a facility is placed in service). ITC eligible taxpayers are fully exempt from the PTC, as are those receiving the Section 1603 grant in lieu of tax credits.

Relief from "Haircuts"

Most taxpayers claiming the PTC must reduce the amount of the credit for any grants, tax-exempt bonds, other forms of subsidized energy financing, and other credits (certain open-loop biomass facilities are not subject to this restriction). The ITC and Section 1603 grant is reduced only for grants received when grants are not taxed as income. Under ARRA, the ITC does not need to be reduced for projects receiving subsidized financing. The Section 1603 grant is also not reduced for subsidized financing.

Relaxation of Owner-Operator and Power Sales Requirements

The PTC requires that energy must be produced by the taxpayer (the owner must also be the operator). Further, under the PTC, power must be sold to an unrelated party to qualify for the credit. This requirement discouraged the use of sale-leaseback or inverted lease structures by PTC-eligible projects. The 1603 grant in lieu of tax credit is available to developers engaged in sale-leaseback or inverted lease structures.[66] Additionally, tax-exempt entities may choose to lease renewable energy facilities from taxable owners able to receive the grant. ITC and Section 1603 grant projects are not subject to the power sales requirement, benefitting behind-the-meter projects.

No Limitation for Passive Investors

Passive investors, or investors not playing an active role in the business, can only use tax credits earned from passive investments to offset other passive income. Section 1603 cash grant recipients are not subject to passive credit limitations.

Allow PTC-Property to Claim the ITC

A second option would be to allow PTC-eligible properties to continue to claim the ITC. If the grant program construction deadline of December 31, 2011, is not extended, certain facilities, primarily wind, open and closed-loop biomass, and geothermal energy facilities, will no longer be eligible for the ITC or Section 1603 grant in lieu.[67] Allowing these types of projects to receive the ITC in lieu of the PTC would overcome some of the PTC limitations discussed above.

There are potential drawbacks, however, associated with allowing current PTC-eligible projects to continue to qualify for the ITC. Investment tax incentives for wind and certain other renewable energy facilities were introduced in the late 1970s.[68] The initial renewable ITC was allowed to expire in the 1980s, but was later replaced with a PTC in 1992. Congress elected to adopt a renewable energy PTC rather than an ITC for two reasons. First, by tying tax credits to production, rather than the cost of capital, the tax incentives were designed to reward the most efficient operations. Second, there were concerns that the ITC for wind of the early 1980s served as a tax shelter, with investors installing inoperable renewable energy equipment generating tax credits to offset income produced from other sources.[69] Allowing PTC eligible projects to be eligible for the ITC creates incentives for investment, rather than performance, and may raise issues similar to those addressed by Congress in the 1990s when choosing to adopt a PTC rather than an ITC.

Investors in renewable energy projects that are not involved in day-to-day operations are generally considered, for tax purposes, to be passive investors. Tax credits or losses of passive investors can only be used to offset other passive income. Renewable energy projects that generate losses in early years of operations may be less attractive to investors because of these passive credit and passive loss rules. The Section 1603 grant was not subject to passive income rules. Investors were able to receive the benefits from the grant immediately, while tax credits oftentimes are carried forward to a number of years until investors generate enough passive income to claim credits earned through passive investments in full.

Exempting renewable energy investors from passive loss rules may increase the attractiveness of renewable energy ventures to investors.[70] Doing so, however, may increase the likelihood that investors would use renewable energy investments as a means to reduce taxable income. Passive loss limitations were adopted in 1986 in an effort to reduce the use of tax shelters.

Alternatives to Enhance Access to Tax Equity

A third option could be to enhance access to tax-equity markets by adopting policies that would facilitate the trading or selling or tax credits.[71] Selling tax credits or trading credits for equity is common practice for developers of Low Income Housing Tax Credit (LIHTC) eligible projects.[72] Developers eligible for the LIHTC have three options: (1) self-shelter (that is, claim the credits themselves); (2) sell the credit directly to an investor; or (3) sell the credit to a syndicator who combines credits across a number of projects. The ability to sell tax credits also allows tax-exempt entities to benefit from investing in tax-credit eligible projects.

There are a number of challenges, however, that would likely arise should Congress consider modifying energy tax credits to function like the LIHTC. First, for example, the size of the market for energy tax credits is smaller than that of the LIHTC. Over the 2009 to 2013 budget window, estimated tax expenditures for low-income housing credits are $30 billion. Tax expenditures on the renewable energy PTC and ITC are estimated to total $7.9 billion between 2009 and 2013.[73] Second, the LIHTC is a permanent feature of the tax code, while the PTC and ITC are currently temporary. Thus, given uncertainty surrounding PTC and ITC extensions, syndicators may choose to avoid energy tax credits.

CONCLUSION

Given the approaching expiration of the PTC for wind, a proposed extension of the Section 1603 grant program may be evaluated jointly with a PTC modification. At the end of 2011, Congress may be asked to evaluate whether the benefits associated with the Section 1603 grant program outweigh the budgetary costs and take appropriate legislative action. This report highlighted the benefit to the renewable energy sector of the Section 1603 grant program, specifically the additional subsidy provided by the grant relative to previously available tax incentives. The added subsidy may lead to additional investments in renewable energy generation capacity, and help to enhance the national renewable energy portfolio. As also noted in this report, however, the added subsidy has added budgetary costs.

Additionally, this report assesses how the renewable energy market responded and performed since the Section 1603 grant program became law in February 2009. As noted in the analysis, motivations and incentives to use the grant program differ among the technologies that qualify for Section 1603 grants. In some cases, some renewable energy sub-sectors have actually experienced market size declines. This observation highlights the complexity of renewable energy markets and indicates that other factors, outside of the Section 1603 cash grant program, can influence market growth. If renewable energy expansion is a key policy objective, Congress may want to consider policies that holistically address these other factors in addition to the availability of tax-based incentives.

Furthermore, the Section 1603 grant program was motivated by difficult economic conditions that resulted in less tax equity investment capacity available for renewable energy projects to realize the value of ITC and PTC incentives. As the economy recovers, Congress might want to assess if Section 1603 is needed to serve its original purpose or if the cash grant options provides the renewable energy sector with additional benefits that stimulate continued market growth.

Finally, this report analyzed the Section 1603 grant program from an economic perspective. Section 1603 grants may be a more economically efficient mechanism than tax credits for delivering benefits to the renewable energy sector. However, subsidies to the renewable energy sector, whether grants or tax credits, are less economically efficient than pricing polluting energy alternatives. Again, a question before Congress is whether the benefits associated with the Section 1603 grant option are worth the budgetary and potential economic costs.

APPENDIX. STATE-LEVEL ANALYSIS OF SECTION 1603 GRANTS

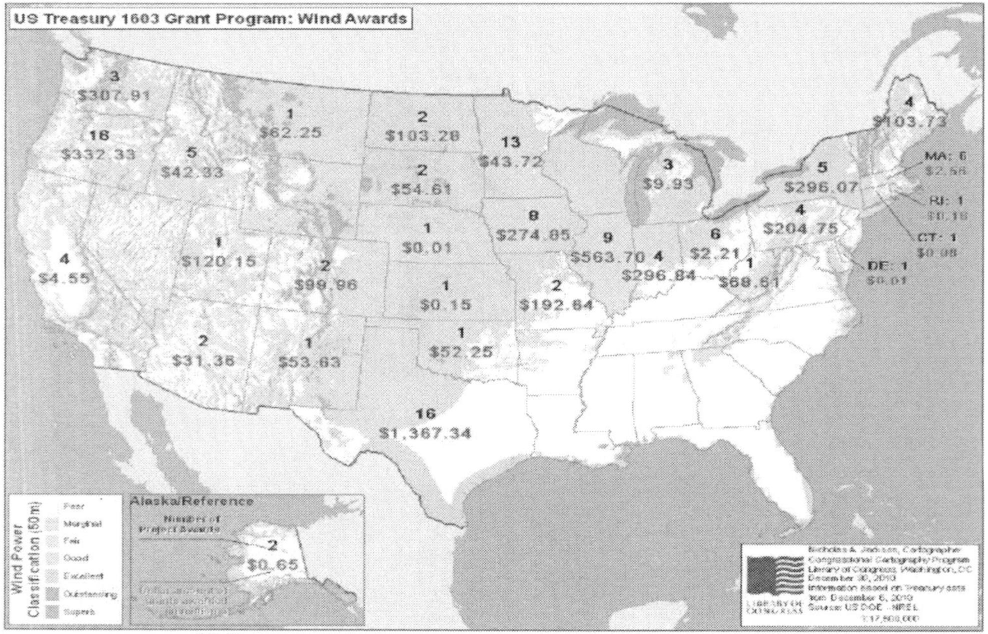

Source: U.S. Department of the Treasury.

Notes: Data used for this map is based on information released by the Department of the Treasury on December 6, 2010. According to data provided by Treasury, Hawaii did not receive any Section 1603 grant awards for wind projects.

Figure A-1. Section 1603 Grant Awards for Wind Projects.

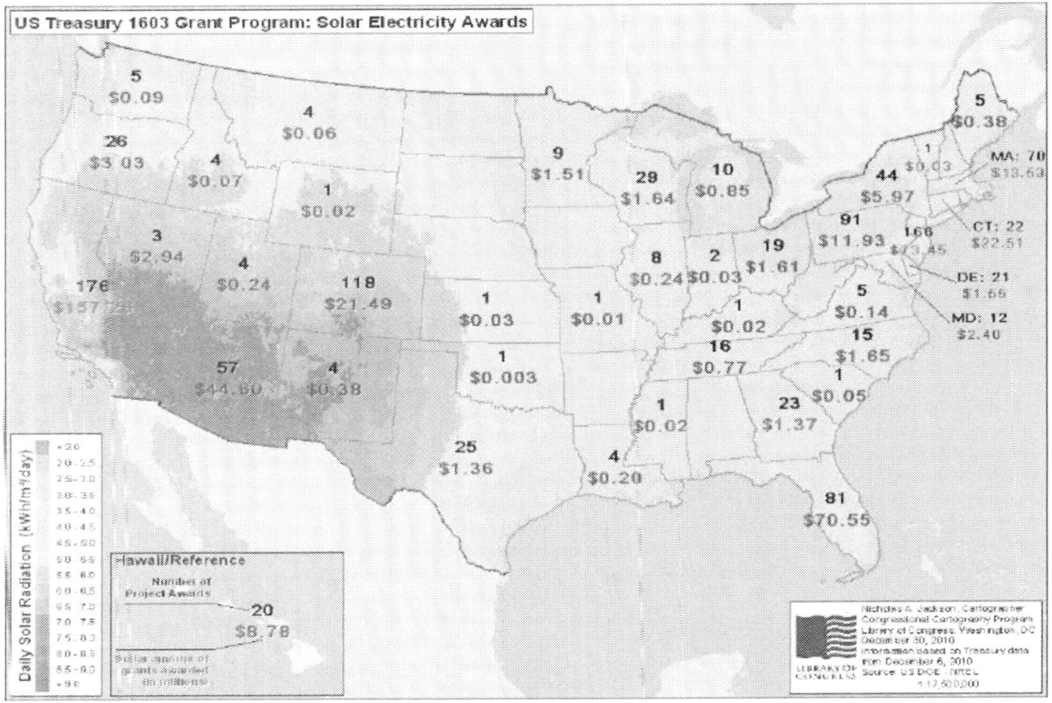

Source U.S. Department of the Treasury.

Notes: Data used for this map is based on information released by the Department of the Treasury on December 6, 2010. According to the data provided by Treasury, Alaska did not receive any Section 1603 grant awards for solar electricity projects.

Figure A-2. Section 1603 Grant Awards for Solar Projects.

ACKNOWLEDGMENTS

Alphonse Anderson, Energy Analyst (Intern), contributed to this report.

End Notes

[1] Energy Information Administration, *State Historical Tables for 2009*, available at http://www.eia.doe.gov/ ... /generation_state.xls.

[2] Treasury guidance provides a detailed description of "beginning of construction" and projects can meet the deadline by demonstrating 1) physical work of a significant nature, 2) construction by contract, or 3) safe harbor (more than 5% of the total cost of the property either paid for or incurred).

[3] IRC § 48 and IRC § 45.

[4] The ITC was introduced as part of the Energy Policy Act of 1978 (P.L. 95-618). A full legislative history of the ITC can be found in U.S. Congress, Senate Committee on the Budget, *Tax Expenditures: Compendium of Background Material on Individual Provisions*, committee print, prepared by Congressional Research Service, 111th Cong., 2nd sess., December 2010, S. Prt. 111-58, pp. 185-190.

[5] The tax credit amount is 1.5-cents per kWh in 1993 dollars, indexed annually for inflation.

[6] The PTC was adopted as part of the Energy Policy Act of 1992 (P.L. 102-486).

[7] For a complete legislative history of the PTC, see U.S. Congress, Senate Committee on the Budget, *Tax Expenditures: Compendium of Background Material on Individual Provisions*, committee print, prepared by Congressional Research Service, 111th Cong., 2nd sess., December 2010, S. Prt. 111-58, pp. 297-204.

[8] For analysis of the effects of the PTC over time, see Ryan Wiser, Mark Bolinger, and Galen Barbose, "Using the Federal Production Tax Credit to Build a Durable Market for Wind Power in the United States," *The Electricity Journal*, vol. 20, no. 9 (November 2007), pp. 77-88.

[9] IRC § 168(e)(3)(B)(vi)(I). Biomass systems are generally classified as seven-year property.

[10] This assumes that taxpayers are able to realize deductions in the year they are eligible to claim them or find some other way to monetize tax benefits. Some renewable projects may not be able to realize the full benefits associated with accelerated or other favorable depreciation schemes if such projects have little taxable income to offset in the early years of operation.

[11] For a general overview of investment subsidies, including investment credits and accelerated depreciation deductions, see CRS Report R41034, *Business Investment and Employment Tax Incentives to Stimulate the Economy*, by Thomas L. Hungerford and Jane G. Gravelle.

[12] Eligible investments include any property with a cost recovery period of 20 years or less.

[13] Mark Bolinger, Ryan Wiser, and Naim Darghouth, "Preliminary evaluation of the Section 1603 treasury grant program for renewable power projects in the United States," *Energy Policy*, vol. 38, no. 11 (November 2010), pp. 6804-6819.

[14] Financial risks include (1) tax capacity needed to realize the value of tax credit incentives; (2) in the case of production tax credits, the performance of the technology being used; (3) ability of projects to find a suitable tax equity investor; among others.

[15] U.S. Treasury Department, *Payments for Specified Energy Property in Lieu of Tax Credits under the American Recovery and Reinvestment Act of 2009: Program Guidance*, March 2010, http://www.treasury.gov/initiatives/recovery/Documents/guidance.pdf.

[16] Mark Bolinger, Ryan Wiser, and Naim Darghouth, "Preliminary Evaluation of the Section 1603 Treasury Grant Program for Renewable Power Projects in the United States," *Energy Policy*, vol. 38, no. 11 (November 2010), pp. 6804-6819.

[17] American Wind Energy Association (AWEA) annual and quarterly reports.

[18] American Wind Energy Association, "Third Quarter 2010 Market report," October 2010, at http://www.awea.org/ documents/reports/2010_third_quarter_report.pdf.

[19] Energy Information Administration, "Monthly Energy Review," December 2010, at http://www.eia.gov/mer/pdf/ pages/sec7_3.pdf.

[20] California Public Utilities Commission, "Market Price Referent," December 1, 2010, at http://www.cpuc.ca.gov/ PUC/energy/Renewables/mpr.

[21] Matthew L. Wald and Tom Zeller Jr., "Cost of Green Power Makes Projects Tougher Sell," *New York Times*, November 7, 2010.

[22] In recent years, bonus depreciation has also been available for investments in wind. The analysis presented below, however, incorporates only the PTC and accelerated deprecation.

[23] For more detail regarding the CRS model used for this analysis, contact Phillip Brown. Phone: 202-707-7386, e-mail

[24] Chadbourne & Parke's April 2010 Project Finance NewsWire states that up to 65% of renewable energy costs are subsidized by the federal government. This statement appears to be based on taking the full value of depreciation benefits provided to renewable projects. CRS analysis, provided in the above figure, estimates the "incremental" value of accelerated deprecation, i.e., MACRS ("incremental value is calculated by subtracting the present value of straight line depreciation from the present value of MACRS depreciation), when compared to straight-line, since MACRS is acceptable for certain energy properties.

[25] Once again, other project parameters (e.g., the required return for tax equity) can potentially alter the incentive type decision. This analysis assumes that all project parameters, except capacity factor and installed cost, are static.

[26] LBNL evaluated a subset of 40 Wind projects that received 1603 cash grants and categorized each project based on the "face value" of the PTC and Cash grant. See "Preliminary Evaluation of the Impact of the Section 1603 Treasury Grant Program on Renewable Energy Deployment in 2009."

[27] Solar Energy Industries Association.

[28] Email correspondence with Treasury officials revealed that some business entities have received awards for multiple projects and 1603 solar awards include approximately 3,700 individual projects.

[29] IRC § 39.

[30] When tax credits are carried forward, previously unused credits are used to offset future tax liability. When tax credits are carried back, taxpayers may request a refund for taxes previously paid, as current tax credits are being used to offset past tax liability.

[31] Shayle Kann, "The Future of the Utility Scale PV in the U.S," *Greentech Media*, December 1, 2010, at http://www.greentechmedia.com/articles/print/the-state

[32] One project categorized as "geothermal electricity" received a $5,000 grant, while the other five projects received grant awards in the range of $28 million to $108 million.

[33] Geothermal Energy Association, http://www.geo-energy.

[34] Treasury defines "open-loop biomass facilities" as follows: "An open-loop biomass facility uses open-loop biomass to produce electricity. Open-loop biomass is any agriculture

livestock waste nutrients or any solid, nonhazardous, cellulosic waste material or any lignin material that is derived from qualified sources." See http://www.ustreas.gov/ recovery/docs/ guidance.pdf.

[35] CRS Report R41440, *Biomass Feedstocks for Biopower: Background and Selected Issues*, by Kelsi Bracmort.

[36] http://www.ustreas.gov/recovery

[37] One particular project listed as "small wind" received $43 million of the $48 million of the small wind awards. Further research indicates that this project includes 132 Megawatts of installed wind generation capacity and utilizes 3- Megawatt Vestas wind turbines.

[38] http://www.ustreas.gov/recovery

[39] Chadbourne & Parke, *Project Finance Newswire,* November 2010.

[40] Chadbourne & Parke, *Project Finance Newswire,* November 2010.

[41] Prior to the availability of Section 1603 cash grants, many wind projects were financed using a "partnership flip" structure and many solar projects were financed using a sale-leaseback structure. Section 1603 allows projects to use a variety of project finance structures to include sale-leaseback (for wind), direct loans, pre-paid PPA, etc. It is likely that financial structures will continue to evolve as the financial community becomes more familiar with cash grants and can structure targeted products for renewable energy projects.

[42] For an overview of various energy market failures and economic rationales for intervention in energy markets, see CRS Report R40999, *Energy Tax Policy: Issues in the 111th Congress*, by Molly F. Sherlock and Donald J. Marples, "Using Tax Expenditures to Achieve Energy Policy Goals," *American Economic Review*, vol. 98, no. 2 (2008), pp. 90-94.

[43] An externality is a spillover from an economic transaction to a third party, one not directly involved in the transaction itself.

[44] Roth and Ambs (2004) use a levelized cost of energy approach that incorporates externalities to evaluate the full cost of generation across 14 generation technologies. They find that externality costs are higher for coal, oil, and gas than for renewables. See Ian F. Roth and Lawrence L. Ambs, "Incorporating Externalities into a Full Cost Approach to Electric Power Generation Life-Cycle Costing," *Energy*, vol. 29 (2004), pp. 2125-2144.

[45] A tax on labor is distortionary as changes in behavior to avoid the tax result in reduced economic activity.

[46] In the energy literature, inframarginal beneficiaries are often referred to as free riders.

[47] Analysis by Bloomberg New Energy Finance finds that the government can achieve more on a per dollar basis when providing grants as opposed to tax credits. See Ethan Zindler, *Cash is King: Shortcomings of U.S. Tax Credits in Subsidizing Renewables*, Bloomberg New Energy Finance, U.S. Policy - Research Note, January 20, 2010.

[48] For analysis of the cost of energy tax incentives broadly, see CRS Report R41227, *Energy Tax Policy: Historical Perspectives on and Current Status of Energy Tax Expenditures*, by Molly F. Sherlock, and CRS Report R40999, *Energy Tax Policy: Issues in the 111th Congress*, by Molly F. Sherlock and Donald J. Marples.

[49] Ultimately, a project's selection of tax incentive depends on the production capacity of projects. Given the option, one would expect projects to select the tax credit with the higher net present value (NPV). While some projects may not perform as expected, one would still expect the aggregate data to reveal the trend with respect to revenue losses noted above.

[50] See footnote 1 of Table 16-1.

[51] The JCT also scored an extension of the Section 1603 grant program as part of the Ways & Mean Committee's Domestic Manufacturing and Energy Jobs Act of 2010. In this preliminary score, a one-year extension was estimated to cost $4.8 billion over five years,

with the cost falling to $2.7 billion over the 10-year budget window. The reduced cost over time reflects anticipated reductions in tax credit payouts over time as grants are paid up front (see Joint Committee on Taxation's score prepared for the Committee on Ways and Means, available athttp://waysandmeans.house.gov/press/PRArticle.aspx? NewsID= 11294)

[52] U.S. Congress, Joint Committee on Taxation, Estimated Budget Effects of the "Tax Relief, Unemployment Insurance Reauthorization, and Job Creation Act of 2010," as Scheduled for Consideration by the United States Senate, committee print, 111th Cong., December 10, 2010, JCX-54-10, available at http://www.jct.gov/publications.html?func= startdown& id=3715.

[53] Joint Committee on Taxation, JCS-3-10.

[54] For additional background, see CRS Report R40833, *Renewable Energy—A Pathway to Green Jobs?*, by Richard J. Campbell and Linda Levine.

[55] The Section 1603 grant application applicants file with the Treasury does request information on jobs and job creation from grant applications. Inconsistency in these self-reported job creation statistics has prevented the Treasury from releasing these data. Further, even if these data were released, it is unlikely that such job estimates would fully incorporate the grant's impact on job creation at the manufacturing level, and other potential feedback effects.

[56] Mark Bolinger, Ryan Wiser, and Naim Darghouth, "Preliminary Evaluation of the Section 1603 Treasury Grant Program for Renewable Power Projects in the United States," *Energy Policy*, vol. 38, no. 11 (November 2010), pp. 6804-6819.

[57] Bolinger et al. (2010) present data on job creation in terms of "job years." Short-term job estimates are the number of one-year full time equivalent positions created. Long-term jobs are assumed to last for the life of the project, generally 20 to 30 years.

[58] American Wind Energy Association, "Tens of Thousands of Layoffs in American Wind Energy Seen at Stake in Tax Extender Package," press release, December 7, 2010, http://www.awea.org/rn_release_12-07-10.cfm.

[59] Solar Energy Industries Association, *Job Creation from Extending the 1603 Treasury Program*, available at http://www.seia.org/galleries/FactSheets/Factsheet_TGP.pdf.

[60] Ibid.

[61] See Congressional Budget Office, *Estimated Impact of the American Recovery and Reinvestment Act on Employment and Economic Output as of September 2009*, Washington, DC, November 2009; Congressional Budget Office, *Estimated Impact of the American Recovery and Reinvestment Act on Employment and Economic Output From October 2009 Through December 2009*, Washington, DC, February 2010; Congressional Budget Office, *Estimated Impact of the American Recovery and Reinvestment Act on Employment and Economic Output from January 2010 Through March 2010*, Washington , DC, May 2010; and Congressional Budget Office, *Estimated Impact of the American Recovery and Reinvestment Act on Employment and Economic Output From April 2010 Through June 2010*, Washington, DC, August 2010. All available at http://www.cbo.gov /publications/ collections/collections.cfm?collect= 16.

[62] Congressional Budget Office, *Estimated Impact of the American Recovery and Reinvestment Act on Employment and Economic Output from April 2010 Through June 2010*, Washington, DC, August 2010, p. 2.

[63] Mark Bolinger, Ryan Wiser, and Naim Darghouth, "Preliminary Evaluation of the Section 1603 Treasury Grant Program for Renewable Power Projects in the United States," *Energy Policy*, vol. 38, no. 11 (November 2010), pp. 6804-6819.

[64] U.S. Congress, Senate Committee on Finance, Ryan Wiser, *Wind Power and the Production Tax Credit: An Overview of Research Results*, Testimony Prepared for a Hearing on "Clean Energy: From the Margins to the Mainstream," 111th Cong., March 29, 2007.

[65] These benefits are outlined in Mark Bolinger, *Revealing the Hidden Value that the Federal Investment Tax Credit and Treasury Cash Grant Provide to Community Wind Projects*, Lawrence Berkeley National Laboratory, LBNL-2909E, January 2010 and Mark Bolinger, Ryan Wiser, and Naim Darghouth, "Preliminary evaluation of the Section 1603 treasury grant program for renewable power projects in the United States," *Energy Policy*, vol. 38, no. 11 (November 2010), pp. 6804-6819.

[66] For further discussion, see Viva Hammer, "Alternative Energy Gets A Second Wind," *Tax Notes*, November 22, 2010, pp. 895-906.

[67] Other PTC-eligible facilities unable to claim the ITC after December 31, 2010, are landfill gas, municipal solid waste, hydroelectric, and hydrokinetic facilities.

[68] For a history of energy tax policy, see CRS Report R41227, *Energy Tax Policy: Historical Perspectives on and Current Status of Energy Tax Expenditures*, by Molly F. Sherlock.

[69] For further discussion, see Viva Hammer, "Alternative Energy Gets A Second Wind," *Tax Notes*, November 22, 2010, pp. 895-906.

[70] Taxpayers with a working interest in oil and gas are exempt from passive loss rules, allowing credits and losses from such interests to be used to offset other income. The President's FY2011 Budget Proposal seeks to eliminate the passive loss exemption for oil and gas.

[71] An additional option that could be considered here could be an incentive similar to the New Markets Tax Credit (NMTC). For additional background, see CRS Report RL34402, *New Markets Tax Credit: An Introduction*, by Donald J. Marples.

[72] For background on the LIHTC, see CRS Report RS22389, *An Introduction to the Design of the Low-Income Housing Tax Credit*, by Mark P. Keightley.

[73] U.S. Congress, Joint Committee on Taxation, *Estimates of Federal Tax Expenditures for Fiscal Years 2009-2013*, committee print, 111th Cong., 2nd sess., January 11, 2010, JCS-1-10.

In: Federal Incentives for Renewable Energy
Editors: L.Belisario and N. Meier
ISBN: 978-1-61324-946-8

Chapter 4

BIOFUELS INCENTIVES: A SUMMARY OF FEDERAL PROGRAMS*

Brent D. Yacobucci

SUMMARY

With recent high energy prices, the passage of major energy legislation in 2005 (P.L. 109-58) and 2007 (P.L. 110-140), and the passage of a farm bill in 2008 (P.L. 110-246), there is ongoing congressional interest in promoting alternatives to petroleum fuels. Biofuels—transportation fuels produced from plants and other organic materials—are of particular interest.

Ethanol and biodiesel, the two most widely used biofuels, receive significant government support under federal law in the form of mandated fuel use, tax incentives, loan and grant programs, and certain regulatory requirements. The 22 programs and provisions listed in this report have been established over the past three decades, and are administered by five separate agencies and departments: Environmental Protection Agency, U.S. Department of Agriculture, Department of Energy, Internal Revenue Service, and Customs and Border Protection. These programs target a variety of beneficiaries, including farmers and rural small businesses, biofuel producers, petroleum suppliers, and fuel marketers. Arguably, in prior years the most significant federal programs for biofuels have been

* This is an edited, reformatted and augmented version of a Congressional Research Service publication, CRS Report for Congress R4011, from www.crs.gov, dated December 23, 2010.

tax credits for the production or sale of ethanol and biodiesel. However, with the establishment of the renewable fuel standard (RFS) under P.L. 109-58, Congress has mandated biofuels use; P.L. 110-140 significantly expanded that mandate. In the long term, the mandate may prove even more significant than tax incentives in promoting the use of these fuels.

The 2008 farm bill—The Food, Conservation, and Energy Act of 2008—amended or established various biofuels incentives, including lowering the value of the ethanol excise tax credit, establishing a tax credit for cellulosic biofuel production, extending import duties on fuel ethanol, and establishing several new grant and loan programs.

Several key biofuels incentives had expired or were set to expire (e.g., a tariff on ethanol imported from most countries, as well as tax credits for biodiesel, renewable diesel, and ethanol) before the passage of the Tax Relief, Unemployment Insurance Reauthorization, and Job Creation Act of 2010 (P.L. 111-312). The incentives included in that law have been extended through the end of 2011.

This report outlines federal programs that provide direct or indirect incentives for biofuels. For each program described, the report provides details including administering agency, authorizing statute(s), annual funding, and expiration date. The **Appendix** provides summary information in a table format.

INTRODUCTION

With recent high energy prices, the passage of the Energy Policy Act of 2005 (P.L. 109-58) and the Energy Independence and Security Act of 2007 (P.L. 110-140), and the passage of the 2008 farm bill (P.L. 110-246), there is ongoing congressional interest in promoting greater use of alternatives to petroleum fuels. Biofuels—transportation fuels produced from plants and other organic materials—are of particular interest. Ethanol and biodiesel, the two most widely used biofuels, receive significant federal support in the form of tax incentives, loan and grant programs, and regulatory programs. The 2008 farm bill also modified existing incentives— including ethanol tax credits and import duties—and established a new tax credit for cellulosic biofuels. The farm bill also authorized new biofuels loan and grant programs, some of which were funded in the FY2010 appropriations cycle.

This report outlines 22 current, expired, or pending federal programs that provide direct or indirect incentives for biofuels. The programs are grouped below by administering agency. The incentives for biofuels are summarized in the Appendix. This information is compiled from authorizing statutes, committee reports, and Administration budget request documents.

ENVIRONMENTAL PROTECTION AGENCY (EPA) RENEWABLE FUEL STANDARD

- Administered by: EPA
- Established: 2005 by the Energy Policy Act of 2005, §1501 (P.L. 109-58); expanded by the Energy Independence and Security Act of 2007, §202 (P.L. 110- 140)
- Scheduled termination: None
- Description: The Energy Policy Act of 2005 established a renewable fuel standard (RFS) for automotive fuels. The RFS was expanded by the Energy Independence and Security Act of 2007. The RFS requires the blending of renewable fuels (including ethanol and biodiesel) in transportation fuel. In 2010, fuel suppliers were required to include 12.95 billion gallons of renewable fuel in the national transportation fuel supply; this requirement increases annually to 36 billion gallons in 2022. The expanded RFS also specifically mandates the use of "advanced biofuels"—fuels produced from non-corn feedstocks and with 50% lower lifecycle greenhouse gas emissions than petroleum fuel—starting in 2009. Of the 36 billion gallons required in 2022, at least 21 billion gallons must be advanced biofuel. There are also specific quotas for cellulosic biofuels and for biomass-based diesel fuel. On May 1, 2007, EPA issued a final rule on the original RFS program detailing compliance standards for fuel suppliers, as well as a system to trade renewable fuel credits between suppliers. On March 26, 2010, EPA issued final rules for the expanded program (RFS2), including lifecycle analysis methods necessary to categorize fuels as advanced biofuels, and new rules for credit verification and trading. While this program is not a direct subsidy for the construction of biofuels plants, the guaranteed market created by the renewable fuel standard is expected to stimulate growth of the biofuels industry and to raise prices above where they would have been in the absence of the mandate.
- For more information: EPA Website, Renewable Fuel Standard (RFS) http://www.epa.gov/otaq/fuels/renewablefuels/index.htm

INTERNAL REVENUE SERVICE (IRS)

Various tax credits and other incentives are available for the production, blending, and/or sale of biofuels and biofuel blends. Tax credits vary by the type of fuel and the size of the producer. Before the enactment of the Energy Improvement and Extension Act of 2008 (P.L. 110-343, Division B), some of the credits allowed taxpayers to blend biofuels produced outside the United States with conventional fuels, export the blended fuel, and claim the tax credit. Section 203 of P.L. 110-343 effectively eliminated this so-called "splash-and-dash" practice by requiring that any fuel eligible for the credit must be produced and/or used within the United States. Recently, concerns have been raised over whether to continue tax credits for fuel that is produced in the United States but is ultimately exported.

Volumetric Ethanol Excise Tax Credit

- Administered by: Internal Revenue Service
- Established: 2005 by the American Jobs Creation Act of 2004, §301 (P.L. 108- 357); modified by the Food, Conservation, and Energy Act of 2008, §15331 (P.L. 110-246); further amended by the Energy Improvement and Extension Act of 2008 (P.L. 110-343, Division B), §203; extended by the Tax Relief, Unemployment Insurance Reauthorization, and Job Creation Act of 2010 (P.L. 111-312), §708
- Scheduled termination: December 31, 2011
- Description: Gasoline suppliers who blend ethanol with gasoline are eligible for a tax credit of 45 cents per gallon of ethanol.
- Qualified applicant: Blenders of gasohol (i.e., gasoline suppliers and marketers)
- For more information: IRS Publication 510, Chapter 2: Fuel Tax Credits and Refunds http://www.irs.gov/publications/p510/ch02.html

Small Ethanol Producer Credit

- Administered by: Internal Revenue Service
- Established: 1990 by the Omnibus Budget Reconciliation Act of 1990, §11502 (P.L. 101-508); extended by the American Jobs

Creation Act of 2004, §301 (P.L. 108-357); expanded by the Energy Policy Act of 2005, §1347 (P.L. 109-58); amended by the Energy Improvement and Extension Act of 2008 (P.L. 110-343, Division B), §203; extended by the Tax Relief, Unemployment Insurance Reauthorization, and Job Creation Act of 2010 (P.L. 111-312), §708

- Scheduled termination: December 31, 2011
- Description: The small ethanol producer credit is valued at 10 cents per gallon of ethanol produced. The credit may be claimed on the first 15 million gallons of ethanol produced by a small producer in a given year.
- Qualified applicant: Any ethanol producer with production capacity below 60 million gallons per year
- For more information: IRS Publication 510, Chapter 2: Fuel Tax Credits and Refunds http://www.irs.gov/publications/p510/ch02.html

Biodiesel Tax Credit

- Administered by: Internal Revenue Service
- Established: 2005 by the American Jobs Creation Act of 2004, §302 (P.L. 108- 357); extended by the Energy Policy Act of 2005, §1344 (P.L. 109-58); amended by the Energy Improvement and Extension Act of 2008 (P.L. 110-343, Division B), §202-203; extended by the Tax Relief, Unemployment Insurance Reauthorization, and Job Creation Act of 2010 (P.L. 111-312), §701
- Scheduled termination: December 31, 2011
- Description: Biodiesel producers (or producers of diesel/biodiesel blends) can claim a per-gallon tax credit. The credit is valued at $1.00 per gallon. Before amendment by P.L. 110-343, the credit was valued at $1.00 per gallon of "agribiodiesel" (biodiesel produced from virgin agricultural products such as soybean oil or animal fats), or 50 cents per gallon of biodiesel produced from previously used agricultural products (e.g., recycled fryer grease). The tax credit had expired at the end of 2009 and was not extended until the passage of P.L. 111-312, which retroactively applies the extension to fuel produced in 2010.
- Qualified applicant: Biodiesel producers and blenders
- For more information: IRS Publication 510, Chapter 2: Fuel Tax Credits and Refunds http://www.irs.gov/publications/p510/ch02.html

Small Agri-Biodiesel Producer Credit

- Administered by: Internal Revenue Service
- Established: 2005 by the Energy Policy Act of 2005, §1345 (P.L. 109-58); amended by the Energy Improvement and Extension Act of 2008 (P.L. 110-343, Division B), §202-203; extended by the Tax Relief, Unemployment Insurance Reauthorization, and Job Creation Act of 2010 (P.L. 111-312), §701
- Scheduled termination: December 31, 2011
- Description: The small agri-biodiesel producer credit is valued at 10 cents per gallon of "agri-biodiesel" (see Biodiesel Tax Credit, above) produced. The credit may be claimed on the first 15 million gallons of ethanol produced by a small producer in a given year. The tax credit had expired at the end of 2009 and was not extended until the passage of P.L. 111-312, which retroactively applies the extension to fuel produced in 2010.
- Qualified applicant: Any agri-biodiesel producer with production capacity below 60 million gallons per year
- For more information: IRS Publication 510, Chapter 2: Fuel Tax Credits and Refunds http://www.irs.gov/publications/p510/ch02.html

Renewable Diesel Tax Credit

- Administered by: Internal Revenue Service
- Established: 2005 by the Energy Policy Act of 2005, §1346 (P.L. 109-58); amended by the Energy Improvement and Extension Act of 2008 (P.L. 110-343, Division B), §202-203; extended by the Tax Relief, Unemployment Insurance Reauthorization, and Job Creation Act of 2010 (P.L. 111-312), §701
- Scheduled termination: December 31, 2011
- Description: Producers of biomass-based diesel fuel (or producers of diesel/renewable biodiesel blends) can claim $1.00 per gallon tax credit. Renewable diesel is similar to biodiesel, but it is produced through different processes and thus is ineligible for the (above) biodiesel credits. The tax credit had expired at the end of 2009 and was not extended until the passage of P.L. 111-312, which retroactively applies the extension to fuel produced in 2010.
- Qualified applicant: Renewable diesel producers and blenders

- For more information: IRS Publication 510, Chapter 2: Fuel Tax Credits and Refunds http://www.irs.gov/publications/p510/ch02.html

Credit for Production of Cellulosic Biofuel

- Administered by: Internal Revenue Service
- Established: January 1, 2009, by the Food, Conservation, and Energy Act of 2008, §15321 (P.L. 110-246)
- Scheduled termination: December 31, 2012
- Description: Producers of cellulosic biofuel can claim $1.01 per gallon tax credit. For producers of cellulosic ethanol, the value of the credit is reduced by the amount of the volumetric ethanol excise tax credit and the small ethanol producer credit (see above)—currently, the value is 46 cents per gallon. The credit applies to fuel produced after December 31, 2008.
- Qualified applicant: Cellulosic biofuel producers
- Note: The credit for cellulosic ethanol varies with other ethanol credits such that the total *combined* value of all credits is $1.01 per gallon. As the volumetric ethanol excise tax credit and/or the small ethanol producer credits decrease, the per-gallon credit for cellulosic ethanol production increases by the same amount.

Special Depreciation Allowance for Cellulosic Biofuel Plant Property

- Administered by: Internal Revenue Service
- Established: 2006 by the Tax Relief and Health Care Act of 2006, §209 (P.L. 109- 432); amended by the Energy Improvement and Extension Act of 2008 (P.L. 110- 343, Division B), §201
- Scheduled termination: December 31, 2012
- Description: A taxpayer may take a depreciation deduction of 50% of the adjusted basis of a new cellulosic biofuel plant in the year it is put in service. Any portion of the cost financed through tax-exempt bonds is exempted from the depreciation allowance. Before amendment by P.L. 110-343, the accelerated depreciation applied only to cellulosic ethanol plants that break down cellulose through enzymatic processes the amended provision applies to all cellulosic biofuel plants.

- Qualified applicant: Any cellulosic ethanol plant acquired after December 20, 2006, and placed in service before January 1, 2013. Any plant that had a binding contract for acquisition before December 20, 2006, does not qualify.
- For information: See Senate Finance Committee, Summary of House-Senate Agreement on Tax, Trade, Health, and Other Provisions, December 7, 2006.

Alternative Fuel Station Credit

- Administered by: Internal Revenue Service
- Established: 2005 by the Energy Policy Act of 2005 §1342 (P.L. 109-58); extended by the Energy Improvement and Extension Act of 2008, §207 (P.L. 110- 343, Division B); expanded by the American Recovery and Reinvestment Act, §1123 (P.L. 111-5); extended by the Tax Relief, Unemployment Insurance Reauthorization, and Job Creation Act of 2010 (P.L. 111-312), §711
- Scheduled termination: December 31, 2011
- Description: A taxpayer may take a 50% credit for the installation of alternative fuel infrastructure, up to $50,000, including E85 (85% ethanol and 15% gasoline) infrastructure. Residential installations qualify for a $2,000 credit (biofuels pumps are not generally installed in residential applications)
- Qualified applicant: Individual or business that installs alternative fuel infrastructure

Department of Agriculture (USDA)

Biorefinery Assistance

- Administered by: Rural Business-Cooperative Service (RBS)
- Annual funding: $74 million in mandatory spending for FY2009, $245 million for FY2010; authorization of and additional $150 million annually for FY2009- FY2012
- FY2010 Appropriation: $245 million to guarantee $691 million in loans

- Established: 2008 by the Food, Conservation, and Energy Act of 2008, §9001 (P.L. 110-246)
- Scheduled termination: End of FY2012
- Description: Grants to biorefineries that use renewable biomass to reduce or eliminate fossil fuel use.
- Qualified applicant: Biorefineries in existence at the date of enactment.
- For more information: See RBS websitehttp://www.rurdev.usda. gov/rbs/busp/ baplg9003.htm

Repowering Assistance

- Administered by: RBS
- Annual funding: $35 million in mandatory funding for FY2009, to remain available until expended, plus $15 million authorized annually for FY2009 through FY2012
- FY2010 Appropriation: None ($35 million in FY2009)
- Established: 2008 by the Food, Conservation, and Energy Act of 2008, §9001 (P.L. 110-246)
- Scheduled termination: End of FY2012
- Description: Grants to biorefineries that use renewable biomass to reduce or eliminate fossil fuel use. RBS issued a Notice of Funding Availability June 12, 2009 http://www.rurdev.usda. gov/rbs/busp/ 9004%20FR%20NOFA.pdf.
- Qualified applicant: Biorefineries in existence at the date of enactment.
- For more information: See RBS website http://www.rurdev. usda. gov/rbs/busp/ RepoweringAssistance.htm

Bioenergy Program for Advanced Biofuels

- Administered by: RBS
- Annual funding: Mandatory funding of $55 million for FY2009, $55 million for FY2010, $85 million for FY2011, and $105 million for FY2012, plus $25 million authorized annually for FY2009-FY2012
- FY2010 Appropriation: $55 million ($30 million in FY2009)

- Established: 2008 by the Food, Conservation, and Energy Act of 2008, §9001 (P.L. 110-246)
- Scheduled termination: End of FY2012
- Description: Provides payments to producers to support and expand production of advanced biofuels. RBS issued a Notice of Contract Proposal June 12, 2009—
- http://www.rurdev.usda.gov/rbs/busp/NOCP%20FR%209005.pdf
- Qualified applicant: Producer of advanced biofuels
- For more information: See RBS website http://www.rurdev.usda. gov/ rbs/busp/ 9005Biofuels.htm

Feedstock Flexibility Program for Producers of Biofuels (Sugar)

- Administered by: Commodity Credit Corporation (CCC)
- Annual funding: Such sums as necessary are authorized to be appropriated—no appropriation to date
- FY2010 Appropriation: None
- Established: 2008 by the Food, Conservation, and Energy Act of 2008, §9001 (P.L. 110-246)
- Scheduled termination: None
- Description: Authorizes the use of CCC funds to purchase surplus sugar, to ensure the sugar program operates at no-net-cost, to be resold as a biomass feedstock to produce bioenergy.
- Qualified applicant: Producer of biofuels using eligible sugar as a feedstock

Biomass Crop Assistance Program (BCAP)

- Administered by: Farm Service Agency (FSA)
- Annual funding: Commodity payment—depends on the number of applications; through October 19, 2010, approximately $243 million had been paid out to projects in 30 states for FY2009 and FY2010
- Established: 2008 by the Food, Conservation, and Energy Act of 2008, §9001 (P.L. 110-246)
- Scheduled termination: End of FY2012

- Description: Dollar-for-dollar matching payments for collection, harvesting, storage, and transportation (CHST) of biomass to qualified biofuel production facilities (as well as bioenergy or biobased products), up to $45 per ton
- Qualified applicant: Person who delivers eligible biomass to a qualified facility
- For more information: See FSA websit http://www.fsa.usda.gov/ FSA/webapp? area=home&subject=ener&topic=bcap

Rural Energy for America Program (REAP)

- Administered by: RBS
- Annual funding: $255 million total in mandatory spending from FY2009- FY2012; an additional authorization of $25 million annually for FY2009- FY2012
- FY2010 Appropriation: $99.4 million in discretionary and mandatory spending to cover $19.7 million in grants and $388 million in loans ($60 million in FY2009 to cover $30 million in grants and $300 million in loans)
- Established: 2008 by the Food, Conservation, and Energy Act of 2008, §9001 (P.L. 110-246)
- Scheduled termination: End of FY2012
- Description: This program replaced the Renewable Energy Systems and Energy Efficiency Improvements program in the 2002 farm bill. The program provides grants and loans for a variety of rural energy projects, including efficiency improvements and renewable energy projects. Although REAP is not exclusively aimed at biofuels projects, the program could be a significant source of loan funds for such projects.

Biomass Research and Development

- Administered by: National Institute of Food and Agriculture (NIFA)
- Annual funding: $118 million total mandatory spending for FY2009- FY2012; up to $35 million additional discretionary funding annually
- FY2010 Appropriation: $28 million ($20 million in FY2009)

- Established: FY2001 by the Biomass Research and Development Act of 2000, §307 (P.L. 106-224); program extended and mandatory appropriations provided by the Farm Security and Rural Investment Act of 2002, §9008 (P.L. 107-171); program extended and funding authorization expanded by the Energy Policy Act of 2005, §941 (P.L. 109-58); significantly modified by the Food, Conservation and Energy Act of 2008, §9008 (P.L. 110-246)
- Scheduled termination: End of FY2012
- Description: Grants are provided for biomass research, development, and demonstration projects. Eligible projects include ethanol and biodiesel demonstration plants.
- Qualified applicant: Wide range of possible applicants
- For more information: http://www.brdisolutions.com/default.aspx

Other USDA Programs

The following programs within the Rural Business Cooperative Service could possibly be used to assist biofuels producers indirectly:

- Business and Industry (B&I) Guaranteed Loans
- Rural Business Enterprise Grants (RBEG)
- Value-Added Grants
- Rural Economic Development Loan and Grant Programs

Department of Energy (DOE)

Biorefinery Project Grants

- Administered by: Office of Energy Efficiency and Renewable Energy
- Annual funding: Approximately $200 million appropriated annually for the biomass program—not all of this funding will go toward biorefinery project grants
- FY2010 Appropriation: $220 million for entire biomass program
- Established: FY2001 through funding authorized in various statutes
- Scheduled termination: None

- Description: This program provides funds for cooperative biomass research and development for the production of fuels, electric power, chemicals, and other products.
- Qualified applicant: Varies from year to year, depending on program goals in a given year
- For more information: http://www.eere.energy

Loan Guarantees for Ethanol and Commercial Byproducts from Cellulose, Municipal Solid Waste, and Sugar Cane

- Administered by: DOE
- Annual funding: Not specified
- FY2010 Appropriation: $0
- Established: 2005 by the Energy Policy Act of 2005, §§1510, 1511, and 1516 (P.L. 109-58)
- Scheduled termination: Varies, depending on specific program
- Description: The Energy Policy Act of 2005 authorizes several programs to provide loan guarantees for the construction of facilities that produce ethanol and other commercial products from cellulosic material, municipal solid waste, or sugar cane.
- Qualified applicant: Private lending institutions, to guarantee loans for the construction of biofuels plants

DOE Loan Guarantee Program

- Administered by: DOE
- Annual funding: $5.4 million for administrative expenses in FY2008; authority for $51 billion in loan guarantees for energy projects in FY2008, including $10 billion for renewable energy and energy efficiency; $6 billion additional FY2009 appropriation to cover $49 billion in loans to all projects (not just biofuels)
- FY2010 Appropriation: $43 million for administrative expenses—to be offset by loan application fees
- Established: 2005 by the Energy Policy Act of 2005, Title XVII (P.L. 109-58)
- Scheduled termination: Not specified

- Description: Title XVII of the Energy Policy Act of 2005 authorizes DOE to provide loan guarantees for energy projects that reduce air pollutant and greenhouse gas emissions, including biofuels projects.
- Qualified applicant: Private lending institutions, to guarantee loans for clean energy projects.
- For more information: http://www.lgprogram.energy.gov/

Cellulosic Ethanol Reserve Auction

- Administered by: DOE
- Annual funding: $1 billion total authorized for all fiscal years; not more than $100 million may be paid in any given year
- FY2010 Appropriation: None; $5 million in FY2008 for the administrative expenses
- Established: 2005 by the Energy Policy Act of 2005, §942 (P.L.109-58)
- Scheduled termination: Not specified
- Description: Section 942 of the Energy Policy Act of 2005 authorizes DOE to provide per-gallon incentive payments for cellulosic biofuels until annual U.S. production reaches 1 billion gallons or 2015, whichever is earlier. DOE finalized regulations on October 15, 2009. http://www.epa.gov/fedrgstr/EPA2009/October/Day-15/i24778.htm.
- Qualified applicant: Any U.S. cellulosic biofuel production facility that meets applicable requirements.

U.S. Customs and Border Protection (CBP) Import Duty for Fuel Ethanol

- Administered by: U.S. Customs and Border Protection
- Annual funding: N/A
- Established: 1980 by the Omnibus Reconciliation Act of 1980 (P.L. 96-499); amended by the Tax Reform Act of 1986, §423 (P.L. 99-514) extended by the Tax Relief and Health Care Act of 2006, §302 (P.L. 109-432); further extended by the Food, Conservation, and Energy Act of 2008, §15333 (P.L. 110-246), and the Tax Relief,

Unemployment Insurance Reauthorization, and Job Creation Act of 2010 (P.L. 111-312), §708

- Scheduled termination: December 31, 2011
- Description: A 2.5% ad valorem tariff and a most-favored-nation duty of $0.54 per gallon of ethanol (for fuel use) applies to imports into the United States from most countries; most ethanol from Caribbean Basin Initiative (CBI) countries may be imported duty-free.
- Covered Entities: Fuel ethanol importers
- For more information: CRS Report RS21930, *Ethanol Imports and the Caribbean Basin Initiative (CBI)*, by Brent D. Yacobucci; Senate Finance Committee, Summary of House-Senate Agreement on Tax, Trade, Health, and Other Provisions, December 7, 2006.

Department of Transportation (DOT) Manufacturing Incentive for Flexible Fuel Vehicles

- Administered by: National Highway Traffic Safety Administration
- Annual funding: N/A
- Established: 1975 by the Energy Policy and Conservation Act of 1975 (P.L. 94- 163); amended by various statutes, most recently the Energy Independence and Security Act of 2007, §109 (P.L. 110-140)
- Scheduled termination: After model year 2019
- Description: Automakers are required to meet Corporate Average Fuel Economy (CAFE) standards for their passenger cars and light trucks. Manufacturers may gain credits for the sale of alternative fuel vehicles, including ethanol/gasoline flexible fuel vehicles (FFVs). However, the credits are limited—the maximum fuel economy increase allowed through the use of these credits is 1.2 miles per gallon through model year (MY) 2014. The credits are phased out after MY2014 and are completely eliminated after MY2019.

APPENDIX. SUMMARY OF FEDERAL INCENTIVES PROMOTING BIOFUELS

Table A-1. Federal Biofuels Incentives by Agency

Administering Agency	Program	Description	Original Authorizing Legislation	FY2010 Appropriation	Expiration Date
Environmental Protection Agency	Renewable Fuel Standard	Mandated use of renewable fuel in gasoline: 4.0 billion gallons in 2006, increasing to 36 billion gallons in 2022	P.L. 109-58 §1501	N/A	None
Internal Revenue Service	Volumetric Ethanol Excise Tax Credit	Gasoline suppliers who blend ethanol with gasoline are eligible for a tax credit of 45 cents per gallon of ethanol	P.L. 108-357 §301	N/A	End of 2011
	Small Ethanol Producer Credit	An ethanol producer with less than 60 million gallons per year in production capacity may claim a credit of 10 cents per gallon on the first 15 million gallons produced in a year	P.L. 101-508	N/A	End of 2011
	Biodiesel Tax Credit	Producers of biodiesel or diesel/biodiesel blends may claim a tax credit of $1.00 per gallon of biodiesel.	P.L. 108-357	N/A	End of 2011

Administering Agency	Program	Description	Original Authorizing Legislation	FY2010 Appropriation	Expiration Date
	Small Agri-Biodiesel Producer Credit	An agri-biodiesel (produced from virgin agricultural products) producer with less than 60 million gallons per year in production capacity may claim a credit of 10 cents per gallon on the first 15 million gallons produced in a year	P.L. 109-58	N/A	End of 2011
	Renewable Diesel Tax Credit	Producers of renewable diesel (similar to biodiesel, but produced through a different process) may claim a tax credit of $1.00 per gallon of renewable diesel	P.L. 109-58	N/A	End of 2011
	Credit for Production of Cellulosic Biofuel	Producers of cellulosic biofuel may claim a tax credit of $1.01 per gallon. For cellulosic ethanol producers, the value of the production tax credit is reduced by the value of the volumetric ethanol excise tax credit and the small ethanol producer credit—the credit is currently valued at 46 cents per gallon. The credit applies to fuel produced after December 31, 2008.	P.L. 110-246	N/A	End of 2012

Table A-1. (Continued).

Administering Agency	Program	Description	Original Authorizing Legislation	FY2010 Appropriation	Expiration Date
	Special Depreciation Allowance for Cellulosic Biofuel Plant Property	Plants producing cellulosic biofuels may take a 50% depreciation allowance in the first year of operation, subject to certain restrictions	P.L. 109-432	N/A	End of 2012
	Alternative Fueling Station Credit	A credit of up to $50,000 is available for the installation of alternative fuel pumps, including E85 (85% ethanol and 15% gasoline)	P.L. 109-58 §1342	N/A	End of 2011
Department of Agriculture	Biorefinery Assistance	Loan guarantees and grants for the construction and retrofitting of biorefineries to produce advanced biofuels	P.L. 110-246 §9001	$245 million to guarantee $691 million in loans	End of FY2012
	Repowering Assistance	Grants to biorefineries that use renewable biomass to reduce or eliminate fossil fuel use	P.L. 110-246 §9001	$35 million in FY2009	End of FY2012
	Bioenergy Program for Advanced Biofuels	Provides payments to producers to support and expand production of advanced biofuels	P.L. 110-246 §9001	$55 million	End of FY2012

Administering Agency	Program	Description	Original Authorizing Legislation	FY2010 Appropriation	Expiration Date
	Feedstock Flexibility Program for Producers of Biofuels (Sugar)	Authorizes the use of CCC funds to purchase surplus sugar, to be resold as a biomass feedstock to produce bioenergy	P.L. 110-246 §9001	No appropriation to date	None
	Biomass Crop Assistance Program (BCAP)	Provides financial assistance for biomass crop establishment costs and annual payments for biomass production; also provides payments to assist with costs for biomass collection, harvest, storage, and transportation	P.L. 110-246 §9001	Dollar-for-dollar commodity payment—approximately $244 million paid out for FY2009 and FY2010	End of FY2012
	Rural Energy for America Program (REAP)	Loan guarantees and grants for a wide range of rural energy projects, including biofuels.	P.L. 110-246 §9001	$99.4 million to cover $19.7 million in grants and $388 million in loan guarantees	End of FY2012
	Biomass Research and Development	Grants for biomass research, development, and demonstration projects	P.L. 106-224	$28 million	End of FY2015
Department of Energy	Biorefinery Project Grants	Funds cooperative R&D on biomass for fuels, power, chemicals, and other products	Various statutes	$220 million total for biomass program	None

Table A-1. (Continued).

Administering Agency	Program	Description	Original Authorizing Legislation	FY2010 Appropriation	Expiration Date
	Loan Guarantees for Ethanol and Commercial Byproducts from Various Feedstocks	Several programs of loan guarantees to construct facilities that produce ethanol and other commercial products from cellulosic material, municipal solid waste, and/or sugarcane	P.L. 109-58 §§1510, 1511, and 1516	No appropriation to date	Varies
	DOE Loan Guarantee Program	Loan guarantees for energy projects that reduce air pollutant and greenhouse gas emissions, including biofuels projects	P.L. 109-58 Title XVII	$43 million FY2010 appropriation to be offset by loan fees	None
				Approximately $100 billion in loan authority from FY2008 and FY2009 appropriations; $10 billion in loan authority for renewable energy and energy efficiency	

Administering Agency	Program	Description	Original Authorizing Legislation	FY2010 Appropriation	Expiration Date
	Cellulosic Ethanol Reserve Auction	Authorizes DOE to provide per-gallon payments to cellulosic biofuel producers	P.L. 109-58 §942	No FY2010 appropriation $5 million in FY2008 for administrative expenses	August 8, 2015
U.S. Customs and Border Protection	Import Duty for Fuel Ethanol	All imported ethanol is subject to a 2.5% ad valorem tariff; fuel ethanol is also subject to a most-favored-nation added duty of 54 cents per gallon (with some exceptions)	P.L. 96-499	N/A	End of 2011
Department of Transportation	Flexible Fuel Vehicle Production Incentive	Automakers subject to Corporate Average Fuel Economy (CAFE) standards may accrue credits under that program for the production and sale of alternative fuel vehicles, including ethanol/gasoline flexible fuel vehicles (FFVs)	P.L. 94-163	N/A	Incentive expires after model year 2019

Source: CRS.

INDEX

C

D

G

H

I

J

L

M

N

O

P

Q

R